# THE *SCYLLA* AND *CHARYBDIS* OF STRATEGIC LEADERSHIP

# THE *SCYLLA* AND *CHARYBDIS* OF STRATEGIC LEADERSHIP

J.R. McKay

Copyright © 2008 Her Majesty the Queen, in Right of Canada, as represented by the Minister of National Defence.

Canadian Defence Academy Press
PO Box 17000 Stn Forces
Kingston, Ontario K7K 7B4

Produced for the Canadian Defence Academy Press
by 17 Wing Winnipeg Publishing Office.
WPO30304

Library and Archives Canada Cataloguing in Publication

McKay, J. R. (James R.)
The Scylla and Charybdis of strategic leadership / J.R. McKay.

Issued by: Canadian Defence Academy.
Produced for the Canadian Defence Academy Press by 17 Wing Winnipeg Publishing Office.
Includes bibliographical references.
ISBN 978-0-662-47815-7 (bound) -- ISBN 978-0-662-47816-4 (pbk.)
Cat. no.: D2-219/1-2008E (bound) -- Cat. no.: D2-219/2-2008E (pbk.)

1. Command of troops. 2. Strategy. 3. Leadership. 4. Military art and science--Canada. 5. Military art and science--United States. 6. Canada--Armed Forces. 7. United States--Armed Forces. I. Canadian Defence Academy II.Canada. Canadian Armed Forces. Wing, 17 III. Title.

U163 M44 2008          355.3'3041          C2008-980242-X

Printed in Canada.

1 3 5 7 9 10 8 6 4 2

# ACKNOWLEDGEMENTS

## Acknowledgements

I would like to thank the following people for their assistance, advice and support for this work:

Howard Coombs, for an excellent suggestion, sage advice and for setting the example for others to follow in terms of embodying a team ethos in academia.

Colonel Bernd Horn, Doctor Robert Walker and Doctor Bill Bentley of the Canadian Forces Leadership Institute, for their wise counsel and support.

Doctor Andrew Godefroy, for invaluable conceptual and research advice and academic commiserations.

My Uncle, Jock McKay, for his keen editor's eye.

The commander, members and esteemed colleagues of Roto 0 Operation ARGUS for sharing a strategic experience with me.

Last but never least, my wife Janet and my daughter Ella, for their infinite reserve of patience and support.

# TABLE OF CONTENTS

**List of Figures and Tables** ........................ iii

**Foreword** ........................................ v

**Preface** ......................................... vii

**Introduction** .................................... ix

**Chapter 1**  *Scylla* and *Charybdis* ..................... 1

**Chapter 2**  Strategy in the Western World from Ancient Greece to the late Industrial Revolution ............. 5

**Chapter 3**  Leadership in the Western World from Ancient Greece to the mid-20th Century ................... 15

**Chapter 4**  *Charybdis*: The Swirling Vortex of Business and Social Science Literature ................... 21

    Part 1:  Strategy in the Business Community ................... 22

    Part 2:  Leadership in the Business Community ................... 35

    Part 3:  Strategic Leadership in the Business Community ........... 38

    Part 4:  Summary ...................... 50

**Chapter 5**  *Scylla*: The Five American Armed Services ......53

    Part 1:  Strategy in the Five American Armed Services ................... 53

    Part 2:  Leadership in the Five American Armed Services ................... 64

# TABLE OF CONTENTS

|  |  |  |
|---|---|---|
| | Part 3: | Strategic Leadership in the Five American Armed Services .................70 |
| | Part 4: | Summary ......................81 |
| **Chapter 6** | | Comparing *Scylla* and *Charybdis* ............83 |
| | Part 1: | *Scylla*'s Five Armed Services and *Charybdis*' Whirlpool of Business Thought on Strategy ...................83 |
| | Part 2: | *Scylla*'s Five Armed Services and *Charybdis*' Whirlpool of Business Thought on Leadership ....................84 |
| | Part 3: | *Scylla*'s Five Armed Services and *Charybdis*' Whirlpool of Business Thought on Strategic Leadership ...............85 |
| **Chapter 7** | | *Odysseus*: The Canadian Forces ............87 |
| | Part 1: | Strategy in the Canadian Forces .......88 |
| | Part 2: | Leadership in the Canadian Forces .....96 |
| | Part 3: | Strategic Leadership in the Canadian Forces ................99 |
| | Part 4: | Reconciling the Canadian Forces with the Two Sources of Strategic Leadership Thought ......................105 |
| **Endnotes** | | ....................................109 |

# LIST OF FIGURES AND TABLES

## List of Figures and Tables

| | | |
|---|---|---|
| Figure 1 | Aim, Objectives and Goals | 6 |
| Figure 2 | Ancient Greek Definitions | 7 |
| Table 1 | Levels of War | 11 |
| Figure 3 | Definitions in Ancient Greek and the Modern Vernacular | 14 |
| Figure 4 | Leadership Schools and Theories | 17 |
| Figure 5 | Schools of Strategy | 25 |
| Figure 6 | Internal Variables of an Organization | 30 |
| Figure 7 | External Variables | 32 |
| Table 2 | Schools of Strategy Formulation | 32 |
| Figure 8 | Definitions of Strategy in the Business Community | 34 |
| Figure 9 | Transactional and Transformational Leadership | 35 |
| Figure 10 | Leadership and Management | 37 |
| Figure 11 | Origins of Strategic Leadership | 39 |
| Figure 12 | Ends, Ways and Means in the NMS | 55 |
| Figure 13 | AFDD 2 Description of the Levels of War | 57 |
| Figure 14 | Political and Military Objective and Stratagems | 59 |
| Figure 15 | Definition of Strategy in the Five American Armed Services | 63 |

# LIST OF FIGURES AND TABLES

| | | |
|---|---|---|
| Figure 16 | Command, Leadership and Management | 66 |
| Table 3 | Navy Reading List on Leadership, Management and Strategic Planning | 67 |
| Table 4 | Coast Guard Leadership Competencies | 78 |
| Figure 17 | Conceptual Change from the Ancient Greek to the Modern Vernacular | 83 |
| Table 5 | Comparison of the Five American Armed Services and the Business Community | 85 |
| Figure 18 | National Interest Framework | 89 |
| Figure 19 | Definitions of Strategy in the CF | 96 |
| Figure 20 | Command, Leadership and Management in CF Doctrine | 97 |
| Figure 21 | Institutional Effectiveness | 100 |
| Table 6 | The Business Community, the CF and the Five American Armed Services Compared | 106 |

# FOREWORD

## Foreword

Words have meaning. At times, words have more than one meaning, and this can lead people to assume that the understanding they have in mind for a word matches what others comprehend as the meaning for the same word. As military personnel, we have an obligation to be clear and concise with terminology to ensure that we communicate effectively. The fate of the nation and the lives of Canadian Forces (CF) personnel depend on our ability to do so.

The term 'strategy' presents us with such a challenge. The author of this monograph explores the ancient Greek roots of the term and how it has changed from a term describing a politico-military plan through the ages to a term describing a master plan, the art of planning or a long term policy. The term 'strategic' provides an even greater challenge as a result. This affects the collective understanding of concepts like 'strategic leadership'.

This monograph seeks to examine the two major influences on CF doctrine with regard to 'strategic leadership'. These are the Business Community and its ever-expanding body of literature on the topic and the Five American Armed Services. It is a primer for the concepts associated with both the Business Community's and the American military's thoughts on 'strategic leadership'.

'Strategic leadership' is, at present, a topic that fills up the shelves of the business section of most bookstores. The author points out that there is a host of different definitions and concepts of 'strategic leadership' in use, but broad trends are identifiable. This is not to say that the literature of the Business Community on the topic lacks value, but one must be aware of the trends that exist in the body of literature before making judgements on its utility for the CF.

The same observation could be made about American military doctrine on the topic of 'strategic leadership'. Our American ally has invested significant thought and effort into the issue and has also tried to glean valuable insights from the body of business literature. The Five American Armed Services and the CF differ in that the political and legal frameworks that affect the formulation and implementation

# FOREWORD

of policy and strategy are different from one and other. As a result, a wholesale adoption of American thought on the issue of 'strategic leadership' is neither feasible nor desirable. As the author identified, the CF and Five American Armed Services have reached a similar and appropriate conclusion – 'strategic leadership' is about leading the institution so that when the time comes, forces are available and prepared to fight in the national interest.


Major-General J.P.Y.D. Gosselin
Commander
Canadian Defence Academy

# PREFACE

## Preface

The Canadian Forces Leadership Institute (CFLI) is proud to release *The Scylla and Charybdis of Strategic Leadership,* another publication in its Strategic Leadership Writing Project under the auspices of the Canadian Defence Academy (CDA) Press. Our intention has always been to promote professional development within the Canadian Forces in regards to leadership, as well as to promote a mechanism to educate the public at large on military matters. This volume achieves that aim.

*The Scylla and Charybdis of Strategic Leadership* is a significant addition to the CDA Press collection. It examines strategic leadership within the context of American and Canadian doctrine and practice. Although as the author points out, the political and legal frameworks that affect the formulation and implementation of policy and strategy in both cases are different, both nations have reached concurrence on the goal of strategic leadership – that is, to lead the institution effectively so that when required, armed forces are available and prepared to fight in the national interest.

I believe you will find this book of great interest and value whether you are a military professional, scholar or simply interested in the study of war and conflict. As always, we at CFLI and the CDA Press invite your comment and discussion.

Colonel Bernd Horn
Chairman
Canadian Defence Academy Press

# INTRODUCTION

## Introduction

The concept of strategic leadership only appears to be simple; in reality, it is far more complicated. The reason that it is so complicated is that the roots of the concept – strategy and leadership – are terms that have more than one meaning. This is not merely a case of meaning different things to different groups of people; even within more cohesive groups such as an armed service or a corporation or a school of business, both terms have multiple meanings. The co-existence of many meanings renders any discussion of strategy, leadership and strategic leadership somewhat pointless without defining each term clearly.

Imprecise use of language is the root of the issue here. Both military and civilian audiences alike have altered the definition of strategy over time to use the term to describe rather diffcrent activities from its Ancient Greek roots. Our collective fascination with leadership has produced more definitions and theories than evidence to sustain them. Strategic leadership, as a term, has been a relatively new phenomenon. Nonetheless, it suffers from the definitional ambiguity of both of its root terms as well as significant dichotomy between how it is described in the business world and in military institutions. For the Canadian Forces (CF), there is an added layer of complication. It, due to its small size and need to adapt to changing political and economic conditions, tends to draw upon other organizations' ideas as sources of doctrine. When it comes to the concept of strategic leadership, there are two major sources of ideas: the Five American Armed Services and the international Business Community. This added layer of complication means that the CF is faced with having to sort useful from those less useful elements of doctrine. This exercise should not be undertaken without understanding both the business and American military thought on strategic leadership and its two roots – strategy and leadership.

This monograph has been undertaken for the Canadian Forces Leadership Institute as part of the Strategic Leadership Writing Project. It is intended to provide an introduction to the concept of strategic leadership, to situate the intellectual context of the concept and explain its evolution in recent decades. This exploration has been undertaken to illustrate the sources upon which the CF can draw and has

# INTRODUCTION

drawn upon for its doctrine on strategic leadership. The roots of the concept of strategic leadership can be traced through the broad Business Community and the American military. The term 'Business Community' refers to the cognoscenti of major corporations and supporting academic communities, whereas the term 'American military' refers to the Five Armed Services (Army, Navy, Air Force, Marines and Coast Guard) and supporting academic communities.

Events in both the Business Community and American military community have affected the concept in recent decades and the evolution of the concepts of strategy, leadership and strategic leadership over time need to be explained as a result. It is not the purpose of the monograph to discuss these three topics in detail but rather to identify the broad trends associated with them to set the context for the examination of American military thought and the Business Community's thoughts on such issues. The aim of this work is to provide that explanation and context to provide clarity to those interested in strategic leadership doctrine and the relationship and exchange of concepts between the business and military communities.

This is an ambitious work and as such, it is necessary to provide caveats to manage the expectations of future readers. This monograph is not intended to be an overly critical or a gentle assessment of recent Canadian strategy or military affairs. It addresses that elusive element of Canadian military thought – military strategy – and seeks to assess that in the light of providing a solid foundation for joint leadership doctrine. The evolution of the concept of strategy described within this work is Euro-centric and does not address strategic thought from other parts of the world in any significant way. It also has, for the sake of parsimony, taken licence with the categorization of the two major streams of thought from the Business Community and American military that feed the concept of strategic leadership. Social science has made significant contributions to the overall study of leadership in both camps. However, most studies on strategic leadership have occurred within the Business Community as opposed to the military, and as a result, unless the work described was undertaken at the military's behest, it has been described as part of the civilian realm. Social scientists are more than free to provide rebuttals as they see fit. Lastly, the amount of material related to the

# INTRODUCTION

study of leadership, strategy, and their closely related disciplines, organizational behaviour, strategic management, strategic planning and group psychology is seemingly endless. This work should not be taken as the final word or authority on such topics.

# Chapter 1
## *Scylla and Charybdis*

In Book Twelve of Homer's *Odyssey*, the protagonist, Odysseus, needed to return to the city of Ithaca to make amends to Poseidon, the god of the sea. In order to travel to Ithaca, he had to sail through the narrow Straits of Messina. Resident in the Straits were two monsters named Scylla and Charybdis and the Straits were so narrow that trying to avoid one would draw one perilously close to the other. Homer wrote that the goddess Circe advised Odysseus that:

> In the other direction lie two rocks, one of which rears its sharp peak up to the very sky and is capped by black clouds that never stream away nor leave clear weather round the top, even in summer or at harvest-time. No man on earth could climb to the top of it or even get a foothold on it, not even if he had twenty hands and feet to help him, because the rock is as smooth as if it had been polished. But halfway up the crag there is a murky cavern, facing the West and running down to Erebus, past which, illustrious Odysseus, you will probably steer your ship. Even a strong young bowman could not reach the gaping mouth of the cave with an arrow shot from a ship below.
>
> It is the home of Scylla, the creature with the dreadful bark. It is true that her yelp is no louder that a new-born pup's, but she is a repulsive monster nevertheless. Nobody could look at her with delight, not even a god if he passed that way. She has twelve feet, all dangling in the air, and six long scrawny necks, each ending in a grisly head with triple rows of fangs, set thick and close, and darkly menacing death. Up to her waist she is sunk in the depths of the cave, but her heads protrude from the fearful abyss, and thus she fishes from her own abode, groping greedily

# CHAPTER 1

> around the rock for any dolphins or seals or any of the larger monsters which Amphitrite breeds in the roaring seas. No crew can boast that they ever sailed their ship past Scylla unscathed, for from every blue-prowed vessel she snatches and carries off a man with each of her heads.
>
> The other of the two rocks, Odysseus, is lower, as you will see, and the distance between them is no more than a bowshot. A great fig-tree with luxuriant foliage grows upon the crag, and it is below this that dread Charybdis sucks the dark waters down. Three times a day she spews them up, and three times she swallows them down once more in her horrible way. Heaven keep you from the spot when she does this because not even the Earthshaker could save you from destruction then.[1]

In short, Scylla was the six-headed terror that lived on the cliffs on one side of the Straits. Charybdis was the massive whirlpool capable of engulfing entire ships that lived on the other side of the Straits. Odysseus had to choose to sail closer to one than the other.

This Greek mythology mirrors the plight of the Canadian Forces (CF) when dealing with strategic leadership doctrine. On the one hand, the American military is a five-headed Scylla on the issue of leadership as each of the armed services (Army, Navy, Air Force, Marine Corps and Coast Guard) is responsible for the development of future leaders; each service shares a common definition of strategy. On the other hand, the Business Community (including the supporting academic disciplines associated with business administration and social science) is just like Charybdis, i.e. a vast swirling pool of thought – largely incoherent – on strategy and leadership. The Business Community's body of thought on strategy is far less coherent than its body of thought on leadership, and it would be charitable to describe the latter as coherent. The mythological conundrum of Scylla and Charybdis applies to the two sources of thought. Avoiding one can lead to the embracing of the other.

# CHAPTER 1

The CF, in this metaphor, is Odysseus' ship, having to pass between Scylla and Charybdis. The CF is faced with the 'Doctrine Writer's Conundrum' where the doctrine writer can choose to adopt elements of an ally's doctrine and be pilloried for intellectual sloth and/or plagiarism, or the doctrine writer could write from first principles and be criticized for insufficient research, unnecessary labour and being painfully slow in the development of a sorely needed doctrinal product.[2] On the one hand, they could be accused of being lazy, and on the other, they could be accused of being arrogant and unresponsive. The CF can adopt elements of the American military thought on the concept of strategic leadership; it could sift through the vast quantities of social science and business literature associated with the concept or attempt to craft a hybrid of both approaches. There is a danger, however, with adopting concepts from other organizations and one must consider the unintended consequences of the application of those concepts in the CF, e.g. adopting concepts inconsistent with the CF or Canadian values, or concepts that erode service ethos or healthy civil-military relations.[3] The choice of approach is no small matter as it could influence the generation of future Canadian military strategic leaders, strategy and the professional health of the CF as a whole.

For the sake of clarity, the term 'strategic leadership' must be deconstructed into its roots of 'strategy' and 'leadership' and then the concept of 'strategic leadership' will be discussed in light of the American military establishment's and the Business Community's concepts. It is acknowledged that this is a Newtonian approach, as the concept will be reduced to its component parts for the sake of simplicity; this should not be construed as a rejection of 'New Science' or related concepts but an attempt to communicate clearly.[4] Clarity of communication is best established by starting with the very roots of strategy and leadership.

# Chapter 2

## Strategy in the Western World from Ancient Greece to the late Industrial Revolution

The concept of strategy has a number of connotations that developed over the ages and the context tends to determine which one is used. The imprecise use of what passes for language is the crux of the issue. R.G. Collingwood, a noted British philosopher and historian, once argued that:

> The business of language is to express or explain: if language cannot explain itself, nothing else can explain it; and a technical term in so far as it calls for explanation, is to that extent not language but something else which resembles language in being significant, but differs from it in not being expressive or self-explanatory.[1]

Collingwood's point seems tailor-made for the term 'strategy'. Over time, the concepts that the term has been used to describe have changed such that it has become a 'technical term' and not language. While it seems obvious that language ought to be used to communicate as opposed to obfuscate, this is not always the case. The Oxford Canadian Dictionary (OCD) contains four different definitions of the term 'strategy', and these are not the only ones in use.[2] It should not come as a surprise that the American military and business communities use different definitions of the term and in the case of the latter, more than one definition is often used within the same book or article, let alone the body of literature. There is no general theory of strategy to unify the definitions and concepts of strategy, and the military and Business Community's use of those terms and concepts exist in relative isolation from one and other.[3] This is only partially correct as neither the military nor the Business Community can be considered as closed systems as

# 2 CHAPTER

might be expected in a democracy; the two have interacted significantly since the Second World War.[4]

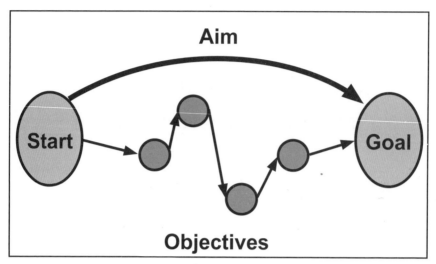

**Figure 1: Aim, Objectives and Goals**

Before discussing the concept of strategy, it is necessary to clarify terminology in advance. Throughout this work, the word 'goal' is the desired end state. For the military community, this represents the political end state and for the Business Community, this represents the organizational end state. The word 'aim' describes the direction toward the goal, and the 'objectives' are the steps that advance one toward the goal. See Figure 1 above for a depiction of all three terms.

Strategy has existed through the ages, but the language to describe it did not always exist. The roots of the term 'strategy' can be traced back to Ancient Greece. In the early Sixth Century BCE, as part of a process of democratic reform, the city-state of Athens reorganized itself into ten tribes. Each of these tribes elected a '*Strategos*' (created from the words *Stratos* 'Army' and *Ago* 'Lead'; plural: *Strategoi*), as its leader and it must be noted that the *Strategoi* held both diplomatic and military responsibilities. The *Strategoi* were leaders and the ten of them formed a war council for Athens.[5] The term *Strategia* (roughly analogous to the term of being an army leader or general ship) was derived from *Strategoi* and became associated with the act of military planning. Another term, *Stratagema*, became associated with specific plans

# CHAPTER 2

or schemes. This term kept the concept separate from examples of its application and therefore contributed to a clear understanding of what described the act of creating a strategy from the product of that act or the individual(s) acting to create a strategy.

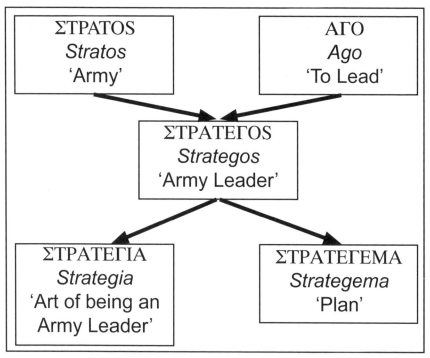

Figure 2: Ancient Greek Definitions

Strategy, as a concept, became an unconscious process during the Dark Ages and remained so until the early modern era. The ancient Greek terms fell out of use. Strategies, strategic thought and practices existed during that time. An example of strategic thought that survived into the Dark Ages and beyond was the Roman military historian Publius Flavius Vegetius Renatus' (henceforth Vegetius) *De Re Militari* was a popular text on military affairs in the Middle Ages and remained popular into the 18th Century.[6] During that same period, military campaigns were conducted to achieve political goals.[7] Although potentially controversial, one could view, in this light, some of the Third to the Fifth Crusades. Forces were assembled and moved to the Middle East to pursue military objectives, namely the destruction of the opponent's source of power

# 2 CHAPTER

in order to achieve the political objective of capturing and retaining possession of the 'Holy Land'. Medieval practitioners and scholars of war would be more than capable of conveying concepts of strategy to others, but they lacked the terms to describe it succinctly. War, from a linguistic perspective, was treated as a monolithic and holistic art. Unfortunately, there were no words available to describe the concept of strategy until much later.[8] The term 'strategy' first appeared during the Renaissance as it was translated from Ancient Greek into French and then from French into English. Yet the Renaissance also saw the beginning of significant political change across the western world.

Over the 17th, 18th and 19th Centuries, most western polities made the transition from kingdoms (based on the principle of the divine right of a monarch to rule) to territorial states (based on rule over a geographically contiguous space) then nation-states (based on rule over a people sharing a common language and culture, ideally in a geographically contiguous space). This process began with the development of the common standard of territorial sovereignty with the Peace of Westphalia (1648). Armies, during the period of territorial states, tended to be small professional armies generated from territorial wealth. They were employed to maintain and/or increase the territory controlled by the ruler; in effect they were being employed to increase the physical span of territorial sovereignty. They tended to fight wars through the skilful use of terrain to enable their manoeuvre to defeat the enemy. As the territorial states began to transform into nation-states and technological advances permitted, armies grew in size and war went from being a duel between two professional bodies into much larger events between nations-in-arms.[9] The definitions of strategy and stratagems blurred. While it was the case that strategy was defined as the art of planning and the stratagem was defined as the product or example of the results of that planning, the term 'strategy' came to encompass both. New stratagems (although they have been described inaccurately as strategies) developed as a result where the defeat of the army was achieved through the rapid destruction of the enemy force (annihilation), the slow destruction of both the force and its logistical support (exhaustion).[10] All of these strategies were intended to bring about a political decision through the force of arms.

# CHAPTER 2

Changes to the practice of strategy over time were mirrored in literature. The oft-quoted Karl von Clausewitz made three major contributions in the field of strategy. First, he provided the language to describe the practice of warfare as either tactics or strategy. This broke the monolithic art of war into two subsets and laid the foundations for strategy to become one of several 'levels of war'.[11] Second, he addressed the issue of contemporary military writing confusing the development and maintenance of an army with its employment in war.[12] The inclusion of methods to raise, train and maintain an army, while vital, obfuscated the understanding of the art of warfare. Modern parlance would have us divide this into force generation/sustainment and force employment. Finally, his definition of strategy fit the emerging paradigm. He argued that strategy was the use of battles to win a war and by so doing achieve a political objective of the state, whereas tactics was the art of winning a battle.[13] This is perhaps his most famous contribution; he argued that the conduct of politics guided warfare and the latter did not occur for its own sake. Military theory and political practice were aligned only in an ideal sense – the reality seldom seemed so clear.

There were other definitions of strategy produced in the 19th Century that better illustrate Clausewitz's argument about politics and warfare. Helmuth von Moltke the Elder defined strategy as: "The practical adaptation of the means placed at a general's disposal to the attainment of the object in view".[14] This definition would see the essence of strategy being the achievement of objectives with the means available to the general in a short timeframe. In some cases, this might mean the entirety of the nation's available resources and in others, only a small portion of its military. The nuance in Moltke's definition reflects the change resulting from the process of democratization; it can be inferred from the definition that politics influence the amount of resources available to the general in question and the amount of time the general has to produce victory. Either way, military strategy was about winning the war using a series of stratagems as quickly as possible whereas policy would dictate when and how warfare would be employed to achieve national goals.[15] Put another way, democratization meant that there was an expectation of rapid success if the use of force was considered.[16] Long, drawn out, and bloody affairs were not something any population would choose deliberately or knowingly.

# 2 CHAPTER

The rise of the nation-states and their democratization changed the nature of strategy significantly over time. While the original concept saw that strategy had both a diplomatic and military facet, strategy by the 20th Century had splintered into a number of different levels. These new levels included 'policy' where a number of different means[17] might to be used to pursue national ends, 'strategy' where military means were associated with political objectives and 'operational art'[18] where campaigns (consisting of a series of battles or engagements) were planned and executed to achieve military ends. Tactics remained the art of winning a battle or engagement.[19] Strategy was less about 'the art of being an Army leader' as Cleisthenes would understand it, than it was about aligning the intentions of these 'Army leaders' with what the government wanted to achieve. Strategy came to be the level at which national political goals were translated into military objectives to which a general or flag officer assigned resources if the state chose to employ military means. Table 1 outlines the four levels of war, the policy and strategic levels, in this document, are the most relevant. As a result, the concept of strategy, in the military sphere, returned to its ancient roots and performs what has been termed a 'bridging function' between the political realm and the conduct of military operations.[20] This 'bridging function' is the source of some confusion as it means strategy is neither policy nor military operations.

The 'bridging' of policy and military operations comes with some risks. Politicians may come to believe that they can delve into the realm of strategy based on the premise that the use of force will achieve political goals easily. Military leaders may come to believe erroneously that success at the tactical and operational levels will guarantee the achievement of policy goals. In western democracies, both the politicians and the military are wary of each other taking on the other's responsibilities.[21] The division of power between the politicians and military means that for the military, as the subordinate organization in a healthy functioning liberal democracy, strategy often takes on an institutional dimension; it becomes the long-range plan to preserve or enhance the military institution in light of potential changes, be they domestically or internationally inspired. Regardless of the institutional dimension of strategy, the concept also came to be associated with a nation-state planning to achieve a political goal through military force to defeat the military force of another nation-state. Yet three connotations of the term (strategy as stratagems; strategy as the translation of political goals into military objectives and the association of military means; and strategy as a long-range plan to preserve or enhance the institution) continued to co-exist.

| Level of War | Activities |
|---|---|
| Policy | Setting of political goals<br>Choice of means to pursue goals |
| Strategic | Translation of political goals into military objectives<br>Selection of means to pursue objectives |
| Operational | Planning of campaigns to achieve military objectives through a series of engagements |
| Tactical | Planning and conduct of engagements |

**Table 1: Levels of War**

A final point has to be made about the military understanding of the concept of strategy that illustrates one of the key differences between the military and business views of strategy. In the Business Community, strategy is a means to ensure the survival of the firm in a very competitive environment. Strategy is seen in a Darwinian light; it is a means to aid if not ensure survival.[22] The military understanding of the concept is different. The French general, André Beaufre, argued that strategy is "...the art of the dialectic of two opposing wills using force to resolve their dispute..."[23] In other words, strategy is a hard fought duel between military forces to resolve a political problem. Such problems exist on a spectrum as different states will resort to the force of arms for different reasons; on one end is the universal motive for the use of force, national survival, and what lies on the other end varies significantly. What differentiates the military and business uses of the term strategy is that in the military use, there is an opponent prepared to use deadly force and risk the effects of the same as opposed to one or more competitors that merely wish to increase their profits. The nature of competition is different in that the stakes are higher and the costs far greater for the military; this leads to greater restraint on the part of most nation-states.

# CHAPTER 2

The Business Community's use of the term strategy has much shallower roots. The concept of strategy, in terms of commerce, can trace its origins to the body of knowledge of management associated with the late industrial revolution. In the early 20th Century, management pioneers, such as Frederick Winslow Taylor and Henri Fayol, sought to develop a scientific approach to industrial enterprise by aligning human efforts to a collective goal. Taylor sought to examine the nature of the work performed in a scheme of mass production with a view to maximizing its efficiency and offered the notion that ensuring a suitable division of labour was the responsibility of managers. Taylor's work represented the application of engineering concepts to human activities. This was based on the assumption that people and machines were alike. If machines could be made to fit organizations, then so too could people. Taylor was not alone in such assumptions and has come to be remembered as the face of a widespread movement called 'Scientific Management'.[24] Fayol was like-minded but instead of trying to find the right people to populate the corporate machine, he sought to optimize the organizations to increase their performance: he attempted to design better organizational machinery to match the capabilities of the people that populated it. Both of them were driven to do so in order to address the increase in the complexity of large business ventures created by the industrial revolution. Scientific management remained in vogue until the late 1930s, when the 'human relations school' of management began to develop.[25] This school of thought was based on philosophical opposition to the mechanistic and potentially authoritarian nature of 'scientific management'. This continued after the Second World War, but the mass mobilization for the war and subsequent demobilization of large numbers of people saw the first wave of military influence on business thought. This influence came from the requirement to populate corporate hierarchies with people capable of leading others, planning ahead and regulating a wide spectrum of activity. Wartime experiences influenced a generation of American business leaders; they saw the value of greater organization, discipline, planning and leadership as opposed to mechanistic or authoritarian approaches.[26] This combination of leadership and planning led to an interest in the concept of strategy within the Business Community.

# CHAPTER 2

It was not until the 1960s that business academics and practitioners wrote about the topic of strategy itself. Notable works included Alfred Chandler's *The Concept of Corporate Strategy* (1962), Alfred Sloan's *My Years with General Motors* (1963) and H. Igor Ansoff's *The New Corporate Strategy* (1965).[27] These works on strategy posited different theories as to what constituted a successful strategy for a corporation, and these theories will be discussed later. It should be noted, however, that the term 'strategy' was applied to an entire corporation as well as to subsets of that corporation's business. The former was labelled as 'corporate strategy' and bears a closer resemblance, in the institutional sense, to the concept of military strategy. Strategies associated with the corporate subsets have come to be known as 'business strategy' and are closer to a series of stratagems employed by different elements of corporations, e.g. marketing, human resource, and so on.

Despite the short time that it has existed as a term used in the civilian world, its definition has changed significantly. The Oxford Canadian Dictionary's definition is illustrative; strategy in the modern vernacular is defined as:

> 1 an esp. long-range policy designed for a particular purpose (economic strategy). 2 the process of planning something or carrying out a plan in a skilful way. 3 a plan or a stratagem. 4a the art of planning and directing military activity in a battle or war...b an instance of this.[28]

Five different connotations of the term exist and four of them are shown in Figure 3 (page 14). Note that the term '*strategia*' has been stripped of its leadership and military criteria in three of the five connotations. A lot of the confusion surrounding the definition of strategy in the Business Community can be attributed to the third definition where the term 'strategy' is interchangeable with 'stratagems', but this does not lessen the effect of some of the connotations of the term strategy, especially with regard to strategy being 'long-range', a planning process or the implementation of a plan. Furthermore, the definitions in Figure 3 refer to both the product (Definitions 1, 3 and 5 (the term stratagem)) and the process (Definitions 2 and 4).

## 2 CHAPTER

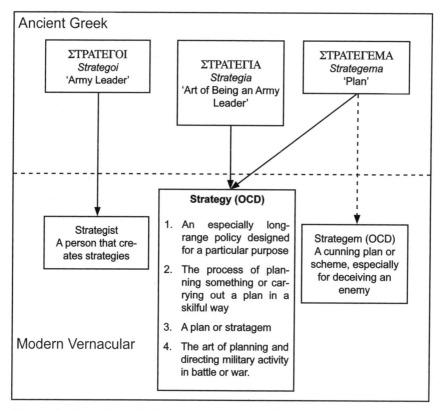

Figure 3: Definitions in Ancient Greek and the Modern Vernacular[29]

To attribute this lack of clarity in the contemporary definitions of strategy purely to its demilitarization would be unfair. The military developed some connotations associated with the descriptor 'strategic'. During the Cold War, the term 'strategic' came to be synonymous with nuclear warfare due to such organizations as Strategic Air Command and the Soviet Strategic Rocket Forces. Since the end of the Cold War, a new connotation has emerged that is more in line with the first definition. The term 'strategic' has come to be considered as not only 'long range' but also holistic in its approach and understanding. To think strategically, for example, is now understood as seeing and understanding the proverbial 'bigger picture' and over a greater span of time.[30] This, however, could be attributed to the adoption of business thought within the ranks of the military.

# Chapter 3

## Leadership in the Western World from Ancient Greece to the mid-20<sup>th</sup> Century

The concept of leadership has evolved throughout the ages, but it was a prisoner of its origins for a significant period of time. As a Western cultural construct, it was based on early theology with regard to hierarchies and social control.[1] Leadership was understood as authority over others. This authority was conferred primarily by 'divine' ordainment, and this meant that those ordained had a 'moral responsibility' to do so.[2] This was the intellectual source of the concept of the Divine Right of monarchs. It was further reinforced by the wealth that rulers would accrue over time and display to their followers as evidence of their worthiness to rule.[3] Over time, the hierarchical nature of society deepened where, by virtue of a contractual agreement, two members of the upper echelons of society would enter into an arrangement of lord and vassal known as feudalism. The latter was a subordinate of the lord, and owed the lord loyalty, aid (including financial) and military service in exchange for military and material support. Both lords and vassals would maintain estates that employed serfs. This created a series of layers between monarchs and those they ruled. This system bound rulers and ruled in a multi-layered political, economic and social hierarchy.

The term 'leader' first appeared in the English language in approximately 1300 AD. It was based on the Old English verb 'leden', which meant 'to guide' or 'to show the way'. It should be noted, however, that the term to describe the art of being a leader came much later.[4] This term 'leader' appeared in language during a period of time where divine right monarchies were beginning to wane.

Economic and technological advances slowly eroded the foundations of the feudal structure by the early modern era. Economic activity had been centred on the

# 3 CHAPTER

collection of rents-in-kind from feudal vassals, but with the widespread adoption of currency and commerce, the payment of rents-in-kind began to lose its importance. This not only broke the economic relationship between ruler and vassal, it also broke the military relationship where feudal rulers would levy military service from the vassals. This coincided with the introduction of gunpowder, which had the effect of making the conduct of warfare available to more than the societal elites. During the early modern era, military service in Europe slowly changed. It went from being a political responsibility to one's superiors in a hierarchy to a source of employment and revenue for the common man.[5] The structures associated with divinely ordained leadership-as-authority supported by a political hierarchy broke down, and the philosophy behind it did as well.

Niccolò Machiavelli has been remembered as the archetype of the cunning planner. One of his contributions was to advocate effective rule in order to achieve political objectives. Leadership was still cast in an authoritarian manner; however, it had changed significantly. Machiavelli noted that leadership was a human activity as opposed to something that required divine intervention.[6] Machiavelli stated that:

> I am not unaware that many have held and hold the opinion that events are controlled by fortune and by God in such a way that the prudence of men cannot modify them, indeed, that men have no influence whatsoever. Because of this, they would conclude that there is no point in sweating over things, but that one should submit to the rulings of chance. This opinion has been more widely held in our own times, because of the great changes and variations, beyond human imagining, which we have experienced and experience every day. Sometimes, when thinking of this, I have myself inclined to this same opinion. None the less, so as not to rule out our free will, I believe that it is probably true that fortune is the arbiter of half the things we do, leaving the other half or so to be controlled by ourselves...I also believe that the one who adapts his policy to the times prospers, and likewise the one whose policy clashes with the demands of the times does not.[7]

# CHAPTER 3

The idea of divine ordainment was irrelevant if rulers could not wield their own authority effectively. They, therefore, needed to take control of their own fortunes to ensure that they remained effective; this control of one's destiny became associated with the notion of heroism.

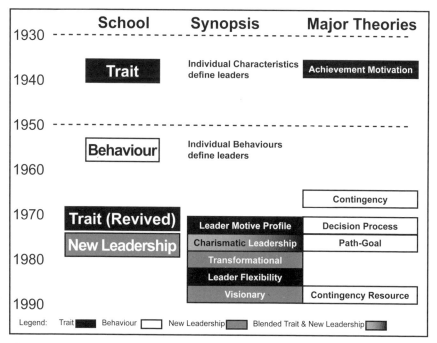

Figure 4: Leadership Schools and Theories

It was not until the turn of the 20th Century that academics began to study the concept of leadership in earnest. This coincided with the development of other academic studies examining related phenomena such as command (through the study of military history or war studies) and management (through the studies of industrial engineering or business administration). Social scientists took an interest in the concept, but not surprisingly, due to the confusion between what actually constituted leadership or management, the term 'leadership' is not defined well or coherently within the social science literature.[8] Most of the leadership literature in the 20th Century dealt with direct (e.g. face to face) leadership by the middle ranks of any organization. The term 'direct leadership' will refer to any form of communication used by leaders to communicate with subordinates in the first person. Conversely, the term 'indirect leadership' will refer to any form of communication used by leaders that does not do so in the first person.[9]

Schools of thought about leadership developed and dominated its study. These schools all existed within the framework of the 'industrial paradigm'; this meant that they tended to be based on Newtonian science, focused on management and managers in particular, goal-oriented, utilitarian, rational and scientific.[10] See Figure 4 (page 17) for a summary. The early 20th Century thought was focused on the centralization and control of activity by leaders, which matched the management thought of that era.[11] From the 1930s to the 1950s, the dominant social scientific approach to leadership was to identify the traits or characteristics of leaders. Its theories included 'Achievement Motivation' (individuals were motivated to perform to attain objectives). This was followed by studies on the behaviour of leaders and these were favoured by a number of management theorists, social psychologists, anthropologists, sociologists, political scientists and even educators.[12] The major theories of 'behaviourism' included 'Contingency' (leaders could use situational control to influence groups), 'Decision Process' (a decision-making aid), 'Path-Goal' (refined version of contingency) and 'Cognitive Resource' (a further refined version of contingency that took into account leader intelligence and experience). Trait-based theories experienced a brief revival in the mid-1970s but other theories soon eclipsed them. The 'New Leadership' school developed in the mid to late 1970s and its major theories are considered to have greater influence at present. They include 'Charismatic Leadership' (leaders as confident, charismatic and assertive), Transformational (use of charismatic leadership to radically alter organizational fortunes), and Visionary (inspiration of followers through the use of symbols and communication).[13]

A final point needs to be made about leadership that applies equally to the CF, the Five American Armed Services, and the Business Community. Regardless of the definitional disputes that follow, it is important to keep in mind that leadership is based on human relationships. Leadership has a *de jure* and a *de facto* dimension and ideally, these two dimensions are congruent. An individual's position in an organization provides the *de jure* dimension of leadership. The individual's ability to influence and motivate others provides the *de facto* dimension. Congruence

# CHAPTER 3

between the two dimensions requires a leader to exercise their authority (positional, moral or otherwise) and others to at least partially recognize and comply with the leader's wishes. Over-reliance on one dimension of leadership will have a deleterious effect. For example, the decline of compliance with leaders' wishes can occur due to the overuse of *de jure* leadership, or the emergence of a different leader within a group due to a lack of authority.[14] People in positions of authority are just that without a subordinates' compliance, but the structures of authority are intended to ensure that, at least initially, compliance is provided.

# Chapter 4

*Charybdis*:
The Swirling Vortex
of Business and
Social Science Literature

As previously stated, Charybdis, in Greek mythology, was a huge and dangerous whirlpool that lay on one side the Strait of Messina. The literature produced for and by the Business Community on strategy, leadership and strategic leadership is a vast swirl of viewpoints on strategic leadership and is daunting like the Whirlpool of Charybdis for it could consume one just in reading, let alone in analysing. The curse of having such a broad array is that is very difficult, if not impossible, to maintain conceptual coherence. It is very possible to become caught up in one or more of the many sub-disciplines, but each must be examined briefly to make sense of the whole.

The Business Community suffers from a lack of definitional consensus with regard to the term 'strategy' but has, over time, developed a loose consensus that leadership is an influence process. This can lead one to conclude that the only issue is the definition of strategy, but this runs the risk of overlooking the importance of the definition of, and theories of, leadership. Strategic leadership does not represent a significant portion of the literature available on leadership, and many authors assume that leadership at the micro-level (i.e. direct) and macro-level (i.e. indirect) are one and the same.[1] The leadership in the literature being discussed suffers from that confusion; strategic leadership is by and large a macro-level process.

# 4 CHAPTER

## Part 1: Strategy in the Business Community

In the first chapter, it stated that the roots of strategy lay in management and the application of science and scientific principles to management. To this list, one might add that the concept of strategy in business has been influenced by economics, sociology and systems analysis.[2] This, however, describes the lineage of ideas as opposed to the philosophical foundations. It is necessary to note that strategy, in business, is about the survival of the corporation or firm in a demanding and unforgiving marketplace. One finds that the language of business is fraught with terminology borrowed from war and economics that has been set in a Darwinian philosophy.[3] Significant failure, inevitably measured in monetary terms, can lead to loss of profits, market share, jobs, or lead to the organization going bankrupt. Survival of the fittest has become the dominant image, and the measure of whether or not an organization survives can be quantified in terms of profit or loss.

Strategy, therefore, is a means or tool to assist business organizations in the battle for survival. It is, at its very essence, about the institution and the first three of the four aforementioned OCD definitions (i.e. a long-range plan for a specific purpose, a planning process or a plan) apply. Unlike the military, where the institution and the polity it serves are different, the institution is the only level of concern. Strategy has been described as both a plan and a posture. The idea of a strategy as a plan is relatively easy to grasp – it represents an intended way of carrying out a task or series of tasks and this corresponds with the third OCD definition. The same cannot be said of a strategy being defined as a posture. This was defined as: 'a relatively stable configuration – a fit or alignment – between mutually supporting organizational elements'.[4] To military minds, this appears to be a desirable state of being for any organization, but this definition leads one to conclude somewhat optimistically that if all the processes are aligned, a strategy exists. This definition of strategy bears no resemblance to the original Ancient Greek concepts and only a passing resemblance to the modern vernacular versions. This notion of 'strategy-as-posture' may not necessarily serve the goal of aiding organizations to survive in a competitive environment. Since that is the purpose of the concept, one should examine how strategy does this. It represents a master plan to 'win', not unlike the definition of the term 'stratagem'. The best

# CHAPTER 4

way of doing so is to ensure consistency, and to support this, strategy can act as a coordinating mechanism for activity in a large organization.[5] In this case, strategy becomes the 'plan of plans' in that it guides the planning of subordinate organizations. It can serve as a means to simplify reality for the organization's members.[6] In this instance, strategy is a heuristic device that provides an image of order and offers a means of attaining objectives in a tumultuous world.

The study of strategy in the Business Community also suffers from a fault line. Some focus on the 'content' of strategy, or what occurs as a result of strategy's application and the means of competition while others focus on the 'process' of how the systems within corporations lead to particular outcomes.[7] One looks at the application of strategy (or stratagems) and the other looks at the development of strategy (or the art of planning and directing activity). There is a risk that evidence gleaned from one is used to deal with issues in the other without making it explicit that the two are at odds. This, above all else, may contribute to confusion within the Business Community with regard to strategy.

The descriptor 'strategic' has been associated with a number of different activities or time horizons in the Business Community. This makes it hard to discern between the activities or horizons, and the term changes with the context of its use, but there is a common thread throughout where the descriptor 'strategic' matches the highest echelons, or pinnacle, of an organization. It has been used to identify the 'upper echelons' of business organizations where 'strategic leadership' meant executive leadership, or the leadership at the organizational pinnacle. 'Strategic' when added to the term management meant policies, processes and decision-making at the organizational pinnacle. When followed by the term 'planning', the definition shifted to being long-range and deliberate. This has a temporal as well as a procedural dimension.

Strategic theories, in business, can be distilled down to two sets of competing philosophies or paradigms[8] with regard to a particular question. This question asks if strategy actually matters to an organization. From this, one can draw the inference that strategic leadership may or may not actually matter as well. While this seems like a nonsensical question, this point is actually the subject

# 4 CHAPTER

of significant debate. On the one hand, choice exists and therefore strategic decisions have an impact. On the other hand, decisions are irrelevant in the face of a competitive and constantly changing environment.[9] The first paradigm will be referred to as the 'Choice Driven Paradigm', which assumes that it is not possible to know the environment completely, but one can make choices about how to operate and survive. The second will be referred to as the 'Environmentally Driven Paradigm', which assumes that it is possible to know the environment and individual choices are irrelevant, as the environment will govern which organizations survive or fail.

Stating that scientific management left a 'rational-instrumental' or mechanistic influence over the concept can summarize the history of the Business Community's use of strategy. Like leadership, it went through a series of trends in the last half of the 20th Century. During this time, the Environmentally Determinist views prevailed, although in recent years, this has eroded somewhat. See Figure 5 for a representation of the schools of strategy.

**Environmentally Determinist Views**

Strategy-Structure-Performance. In the 1960s, Alfred Chandler's early works on strategy led to the emergence of the 'Strategy-Structure-Performance' view.[10] Chandler's studies of a number of successful organizations, such as Dupont, Standard Oil, Sears Roebuck and General Motors led him to conclude that strategy was best used to develop the right organizational structure. Success was dependent on the firm's structure being aligned with its internal characteristics such as the managerial capabilities, the workers and the infrastructure. Adaptation was based on a good fit between an organization's structure and its capabilities. Strategy, in this case, was a means to better organize corporations to mobilize the work force to accept and to align with a common goal, which corresponds roughly to the second OCD definition of strategy (i.e. the process of planning). However, this approach tends to attribute the fortunes of a corporation to its strategy and relate these to the leadership. In essence, Chandler's model tends to lead to 'great man' views of leadership by focusing narrowly on the top management.[11] Leadership's heritage as a form of hero worship also affects the

# CHAPTER 4

Business Community's understanding of strategy in that the top management had become the new *'strategoi'*.

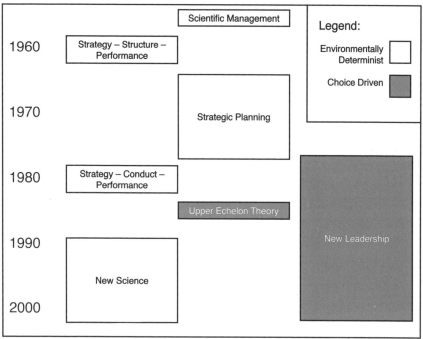

Figure 5: Schools of Strategy[12]

Strategic Planning. This school of strategy was the true intellectual heir of scientific management. Ansoff's *The New Corporate Strategy* was the catalyst for 'strategic planning'.[13] It reached its apex in 1960s and emphasized the importance and value of long-term strategy and control mechanisms.[14] Strategy, in this paradigm, corresponds roughly to the first OCD definition of strategy (i.e. strategy as a long-range plan). The fact that it was a rationally-based school should not be a surprise as the 1960s saw the rise of social scientific methodology across the United States; Secretary of Defense Robert McNamara and his 'whiz kids' represented a manifestation of this school of thought in government. In this school, the concept of strategy represented long-term goal setting, which formed a coordination mechanism only when supported by policies within an organization. Military organizational designs heavily influenced strategic planning's assumptions.[15] If a structure existed, then it was necessary to ensure that it functioned appropriately

# 4 CHAPTER

through coordination. Its similarity to military planning in the era is striking. A military objective would be attained through coordinated plans among numbers of organizations. For example, the United States planned to prosecute nuclear war, should it have become necessary, through the Single Integrated Operating Plan (SIOP).[16] However, this school of thought held that one would have to analyze the environment to arrive at the right structure, mobilize that structure and maintain its focus over time. 'Strategic planning' was based on the assumption that the environment was a knowable entity; this forced organizations to focus on the analysis of all events. However, the notion that the environment was knowable was refuted by events in the 1960s and 1970s, e.g. the Vietnam War, the 1973 oil shock and follow-on energy crises, which signalled the death knell of strategic planning as anything other than a process.[17]

Structure-Conduct-Performance. This represented a blending of a rational approach with environmental determinism. 'Structure-Conduct-Performance' held that if one based the long-range plan on reacting to environmental demands, one ought to be able to maximize profits. The most well known work of this school was Michael Porter's *Competitive Strategy: Techniques for Analyzing Industries and Competitors*.[18] The structure in question was not the corporation but rather the industry in which the corporation was competing. This was a new exercise in environmental determinism. Industry specific variables dictated the success or failure of the firm's conduct and performance. Under this logic, strategy was about ensuring that the environment could be understood and the firm's structure, processes and products were adjusted to meet its challenges, which corresponds roughly to the second OCD definition of strategy. The role of leaders under this paradigm was still minimal.

Agency Theory and the Resource-Based View. Two other theories of strategy merit mention: Agency Theory and the Resource-Based View. Agency Theory was far less influential and it owes its origins to sociology. This one stems from the notion that leaders at all levels make decisions based on self-interest. This can be aligned to organizational interests through the appropriate level of supervision and offers of incentives for better performance.[19] This is another rational view of strategy, although it is different in that it uses the individual as the prime unit of analysis and strategy is less relevant. The Resource-Based View (RBV) is another rational construct, and it represents a throwback to 'Strategy-Structure-Performance'.[20] In RBV, the approach to the term 'resources' is holistic; resources may include the competencies and capacities of ones' organizations, the personnel, the finances, etc. A strategy is intended, under RBV, to mobilize the firm's resources to the level of performance required. The external environment, in turn, sets the performance requirement.[21] In RBV, strategy is a coordination mechanism, which bears little resemblance to the OCD definitions.

'New Science'. This approach to strategy is deceptively simple and seemingly easy to grasp. It is another reaction to the influence of scientific thought on management and it represents a rejection of Newtonian models of science that reduce things to their component parts in favour of general systems theory or complexity theory.[22] In this theory, systems are defined as: "an organized or complex whole: an assemblage or combination of things or parts forming a complex or unitary whole."[23] General systems theory emerged as an organizational theory, but originated from the field of biology.[24] An open system is one that exchanges matter with its surrounding environment like a natural organism.[25] Business organizations fit this definition of an open system. General systems theory also spawned other terms, such as 'systems thinking', which: "...requires conceiving of management dilemmas as arising from within a system with interdependent elements, subsystems, and networks of relationships and patterns of interaction."[26] This involves taking holistic as opposed to reductionist points of view to establish patterns of interaction within a system as opposed to the system's mere components.

Under complexity theory, which is related to general system theory, order is represented by a constant flux as opposed to a stable equilibrium. Change in the environment is therefore constant and end-states do not exist.[27] Patterns define order and not any single state of being. There are two streams of thought within this school. Advocates of the 'New Science' approach argue that organizations ought to mirror the 'self-organizing systems' prevalent in nature. This emphasizes the importance of relationships between people and networks within organizations.[28] This theory provided some intellectual support for the theory of

# 4 CHAPTER

empowerment. This idea of 'self-organizing systems', of course, is reliant on the concept of shared vision.[29] This school of strategy is based on the idea that the external environment of any organization is very complex and difficult for any one individual to sense, let alone understand.[30] Success in this environment is reliant on gathering all available information to make sense of it all. The problem with information-gathering is that no one individual can process it effectively to serve the many needs of the people within an organization; different people need the same information for a number of different purposes.[31] In order to deal with this, this theory of strategy holds that organizations ought to engage in: "...a trade-off of less control for more adaptation through the development of creative self-organizing systems within the organization, again in function of greater flexibility and creativity."[32] This is a trading away of unity of effort to set the conditions to seize opportunities and adapt. In short, it prescribes that organizations ought to seek to relax their control of the conditions to achieve organizational goals. Another view of this would be to advocate the slaving of an organization's hierarchy to existing social networks to achieve success.[33] This theory holds that everyone within an organization recognizes what will lead to success and has the power of choice to pursue it. This does not make a strategy.

**Choice Driven Views**

Upper Echelon Theory. In reaction to the marginalization of the role of leaders and managers, a competing paradigm began to manifest itself in the mid-1980s in the name of 'Structure-Conduct-Performance'. 'Upper Echelon Theory' developed in reaction to the rationalist approach found in that period as exemplified by Michael Porter's *Competitive Strategy*. The proponents of 'Upper Echelon Theory' felt that works like Porter's removed any vestige of managerial or leadership influence over the outcomes.[34] They rebelled against the environmentally determinist paradigm by arguing that managers, and by inference leadership, mattered. 'Upper Echelon Theory' posits that the amount of discretion afforded to 'top managers' governs the degree of impact of their decisions. If they have more latitude, their decisions will have greater impact. Also, the converse was also seen to hold true: With a detailed plan, leaders were less relevant, but the organization would be less capable of adapting to changes.[35] The logic of 'Upper Echelon Theory' applied equally

# CHAPTER 4

to the formulation of strategy. It assumes that 'top managers' are well positioned to make decisions based on their values and the wealth of previous experience. In essence, the theory holds that experience and values allows them to perceive the environment better, thus influencing their decision-making accordingly, and this ought to lead to greater organizational adaptability. 'Upper Echelon Theory' formed the basis of the study of strategic leadership as a whole and represents a study of executive decision-making.[36] The study of strategic leadership in the Business Community is predisposed to the Choice Driven paradigm as a result.

<u>'New Leadership'</u>. This school, which touches both strategy and leadership in the Business Community, is based in part on a rejection of earlier models of organization such as Taylor's scientific management where people were fit into the organizational model as opposed to the reverse.[37] Increased individualism in North American society in recent decades has forced changes to organizations; they have had to accommodate their employees more than they had in the past due to changing cultural norms. The other source of this movement is based on the American Business Community's sense of loss. From the Second World War through to the 1970s, the United States was very prosperous and competition from other nation-states was insufficient to challenge American industry significantly. Since that time, however, the 1973 oil shock, Western European and Japanese gains in international markets created a sense of loss in the 1970s that has lingered.[38] American confidence may have been restored somewhat over the 1980s but was shaken again in the new millennium with a series of ethical crises within the Business Community such as the Enron, WorldCom and Andersen scandals.[39] 'New Leadership' thinking holds that rationalism had led some elements of the Business Community to lose their way in pursuit of profit margins, but proper people-oriented leadership could restore the organizations to the right path. One example of this type of thinking stated that: "...people are the key strategic resource, and strategy must be built on a human-resource foundation."[40] This school of strategy is oriented on the issue of charisma-based leadership of organizations through the use of symbolism and emotional attachment.[41] It can be argued that this is a 'purpose-process-people' doctrine as opposed to 'strategy-structure-performance' or 'strategy-conduct-performance' doctrine'.[42] It includes the 1990s phenomenon of empowerment where subordinates have both the authority and the

# 4 CHAPTER

responsibility to carry out their assigned tasks with minimal interference.[43] The 'New Science' school of strategy and 'New Leadership' have a lot in common. The major assumption at the root of this school is that leaders will perceive and react to environmental changes as needed for the sake of the organization. This, over time, adapts the organization to the environment coherently as opposed to a centrally controlled bureaucracy that resists change.[44] The blending of the Choice Driven and Environmentally Determinist paradigm does not necessarily represent a compromise. Strategy, in this school, is the servant of a charismatic individual that chooses to adapt to the environment as opposed to an organization adapting to its environment just to survive.

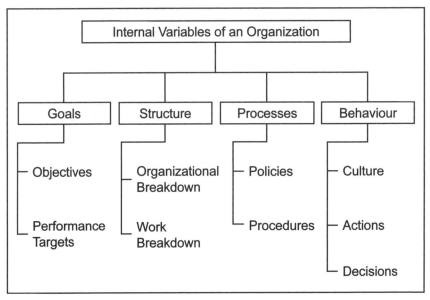

Figure 6: Internal Variables of an Organization

The Business Community's view of strategy has been dominated by an elemental debate that keeps reoccurring through different guises. Some theories hold that strategy matters due to existence of individual choice and therefore decision-making is the key (hence the Choice Driven Paradigm). Other theories hold that the environment will ultimately determine strategies and that individual choice matters far less than identifying trends and understanding the environment (hence the Environmentally Determinist Paradigm).[45] On one hand, strategy (and by inference leadership) is crucial; on the other, it is more or less irrelevant. This clouds the Business Community's use of the concept significantly. If strategy itself does not matter, only stratagems that survive the test of the environment remain relevant. The environmental test is based on whether or not the firm has a net gain or loss in the marketplace. Those that experience too high a degree of net loss cease to survive.

**Dimensions of Strategy**

In the Business Community, the concept of strategy has a number of dimensions, such as its scale, time horizon, factors and phases. The term scale refers to the organizational level of the strategy. The definition of strategy used is dependent on the scale. If the scale is organizational, like in a corporate strategy, then the definition of strategy as a long-term plan is used; if the scale is a subset of an organization, like in a business strategy, then the definition of strategy-as-stratagem is used. This is helpful as it assists in defining other dimensions. The long-term plan naturally has a greater time horizon than the stratagem. The veritable host of business organizations leads to a number of different views on the definition of the term. There is only a broad consensus on the internal variables of any given corporation, and this surrounds four vague terms, which are the goals, the structure, processes and behaviour.[46] See Figure 6 for a depiction of the internal variables.

The external variables have been summarized as the environment, which may include economic aspects such as the market, the industry, and competition, political aspects such as the national and international laws and policies regulating commerce, cultural aspects surrounding the use or abuse of one's products and technology.[47] See Figure 7 (page 32) for a depiction of the external variables. The weighing of the internal and external variables is dependent on one's selection of strategic paradigm. Environmental determinists would argue that external variables govern whether or not the internal variables of an organization will be suitable to allow the organization to survive if not flourish; those advocates of choice would argue that the environment has an effect but not the only effect.

# 4 CHAPTER

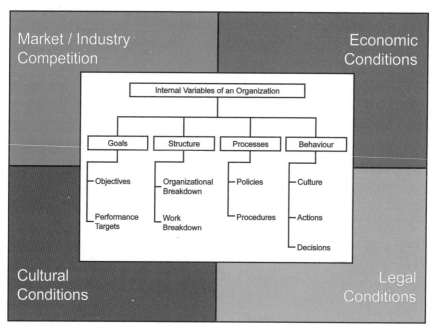

Figure 7: External Variables

| School | Definition |
|---|---|
| Design | A process of conception |
| Planning | A formal process |
| Positioning | An analytical process |
| Entrepreneurial | A visionary process |
| Cognitive | A mental process |
| Learning | An emerging process |
| Power | A negotiation process |
| Cultural | A collective process |
| Environmental | A reactive process |
| Configuration | A transformational process |

Table 2: Schools of Strategy Formulation[48]

# CHAPTER 4

Strategy, regardless of the definition in use, can be broken into three broad phases. These are:

- formulation (defined as the process of developing a strategy)
- implementation (defined as the execution of a strategy)
- control (defined as the actions taken to ensure subordinate elements adhere to the strategy and to adjust those parts of the strategy that are unhelpful or lack utility).[49]

Each phase will be discussed in turn.

Strategy formulation can be a deliberate or an unconscious process, i.e. it may be planned by an organization or formed from custom or practice. Strategy formulation is based on the 'interplay' between the environment, the corporate bureaucracy and the leadership (moderating force). The strategy formulated through the interplay reflects the pattern of consistency.[50] The formulation of strategy does not follow a single codified process. Henry Mintzberg noted that there are many ways or processes to arrive at a strategy, and these represent different 'schools' of strategy formulation as depicted in Table 2.

Of these schools, the 'Learning' process merits further discussion. Most of the processes previously discussed treat strategy formulation as a conscious or deliberate act to shape future behaviour. In some cases, strategies develop from past and/or present behaviour; what were patterns of behaviour become customary and represent an emergent strategy.[51] This is similar to strategy in the 'New Science' view where the organization adapts to its environment. It is not really a strategy but a general acceptance or codification of existing behaviour.

The implementation and the control of strategy, in the literature, have not received as much attention. It appears that most authors focus on how strategies come into existence as opposed to how they are implemented in organizations. There are exceptions to this rule and they come across as glimpses of the blindly obvious. Implementation is often described as 'selling the plan' within an organization in order to minimize resistance to the point of irrelevance. It has been noted

# 4 CHAPTER

that organizational politics often subvert the effective implementation and, as a result, strategies may be built on principles of centralization of control to prevent subversion.[52] This is less of a criticism of the concept of implementation than it is a criticism of the means to mitigate the effects of organizational politics.

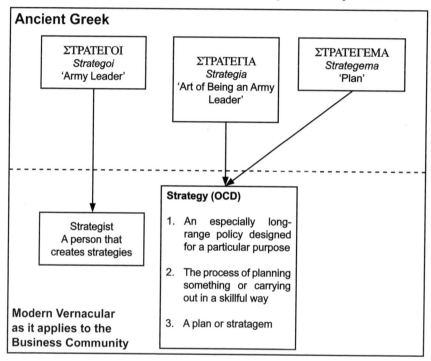

Figure 8: Definitions of Strategy in the Business Community

In summary, strategy in the Business Community suffers from a definitional anarchy that clouds any discussion of the topic. Figure 8 above shows only three definitions of the topic in use and it would not be possible to include them all in a simple diagram. Some definitions in use, e.g. strategy as a posture, bear little resemblance to the Ancient Greek and have been excluded as a result. This definitional anarchy leads to interminable debates over content or process, strategy or stratagems and formulation or implementation. Such debates can be reduced to a central issue – whether or not leadership matters in terms of strategy.

# CHAPTER 4

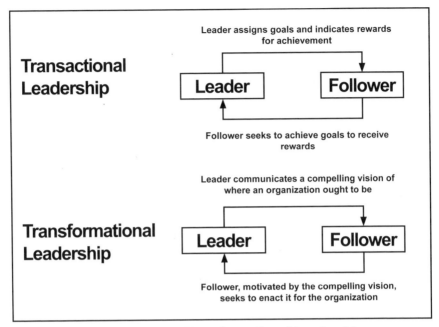

Figure 9: Transactional and Transformational Leadership

## Part 2: Leadership in the Business Community

There is no set definition of the concept of leadership in the Business Community; there is, however, a loose consensus on leadership being an influence process. One must be careful with this definition, as it is philosophically in tune with the 'New Leadership' school of thought.

**'New Leadership'**

There are two major sources of 'New Leadership' thinking. One can be traced to a historian's works on leadership in the late 1970s and to the roots of strategic leadership theory itself. Readers will recognize the first strand as 'Upper Echelon Theory' and its origins as a reaction to the 1970s obsession with analysis. The field of organizational sociology arrived at different conclusions than that of the management field in the late 1970s and early 1980s, and this led to the emergence of 'Upper Echelons Theory'.[53] Readers will also recognize elements of the 'New

# 4 CHAPTER

Leadership'. In 1978, the historian James McGregor Burns published the book *Leadership*.[54] In this book, Burns identified two types of leadership and these are depicted graphically in Figure 9 (page 35). The first was transactional. This would see leaders offer followers rewards for their loyalty and service. It treats every interaction as if it were an economic exchange involving the intangible currencies of leadership and loyalty. The other was transformational leadership. Leaders set and articulate a 'vision' and goals for subordinates to achieve; influenced by the compelling 'vision', the followers set out to achieve those goals.[55] A vision is:

> ...a realistic, credible, attractive future for an organization. It is a carefully formulated statement of intentions that defines a destination or future state of affairs that an individual or group finds particularly desirable. The right vision is an idea so powerful that it literally jump starts the future by calling forth the energies, talents, and resources to make things happen.[56]

Transformational leadership is the philosophical source of empowerment with the exception that the concept of making individual authority align with their level of responsibility has not been included. Burns' concepts took a few years to gain widespread acceptance, but since the mid-1980s, they have become the dominant school of leadership. This domination has meant that the definition of the leadership has become inextricably linked with 'New Leadership'.

The dominance of 'New Leadership' notwithstanding, the concept of leadership has been described in a number of different ways over the years within the literature and the majority of the descriptors have been influence-based.[57] This refers to *de facto* leadership as opposed to its *de jure* version. One memorable variation implied that leaders were, in fact, the servants of their subordinates as they were responsible for: "*...the creation and fulfillment of worthwhile opportunities by honourable means...*"[58] Those opportunities, of course, were intended for subordinates. This represents an important dimension of leadership, especially for those using the concept of empowerment, where leaders set their subordinates up for success.

# CHAPTER 4

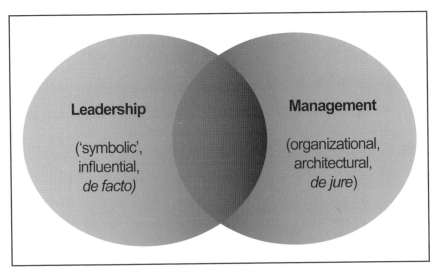

Figure 10: Leadership and Management

The danger with the use of the 'leadership-as-influence' model is that it tends to place management in a negative light. The Business Community's literature tends to describe leadership as a process involving influence and management as a 'control process'.[59] Stephen Covey argued that leadership is about direction (effectiveness) while management deals with speed (efficiency). The 'leadership-as-influence' model holds that leaders motivate, build and influence teams of subordinates while managers organize structures and systems as shown in Figure 10 above.[60] Note that all of the tasks are associated with leaders are described as virtuous and the negative tasks are associated with managers. This, in part, is a reaction to the 'managerial mystique' of the 1970s and an attempt to place the concept firmly in the hands of charismatic as opposed to rational individuals. Such portrayals are misleading, as leaders require both organizational and influential skills.[61] Leadership can be conceived as 'good management' in that it represents a pragmatic, utilitarian, rational and ethical application of basic management principles and techniques.[62] This concept has much merit and should not be overlooked. The portrayal of leaders as charismatic influencers of others opens the concept of leadership to criticism. For example, it has been argued that the concept of leadership is merely a convenient way of explaining complex phenomena associated with organizational fortunes, or in plain English, an unintentional form of hero worship.[63] This, of course, begs the

# 4 CHAPTER

question that if such a criticism is valid and if it is merely a heuristic device, is there a point to the study of strategic leadership? 'New Leadership' has done a disservice to the study of leadership by emphasizing the importance of *de facto* leadership at the expense of *de jure* leadership.

## Part 3: Strategic Leadership in the Business Community

The Business Community's interest in strategic leadership has waxed and waned over the past forty years. During the 1960s and early 1970s, there was a fascination with the role of executives, but this waned as organizational theorists, supported by their rational models and evidence, pointed out that leadership mattered less than had been believed. The results of the Profit Impact of Marketing Strategy (PIMS) Study, a data gathering exercise conducted in the 1970s and 1980s that involved a number of major corporations, suggested that there was little to no impact of leadership on organizational performance.[64] 'Upper Echelon Theory' grew up in reaction to the perceived exclusion of the effects of leadership and management, and combined with the 'New Leadership' school, led to a rebirth in the study of leadership at the organizational level.[65] Since that time, the number of practitioners and salesmen of the art of strategic leadership have proliferated significantly.[66]

**Concept and Definition**

Strategic leadership, as a concept, is focused on the level of entire organizations or corporations. Its philosophical basis, like most of the literature associated with the Business Community, is the survival of the organization in a Darwinian world filled with competition. In terms of scale, strategic leadership focuses on the macroscopic level or that of entire corporations.[67] Most theories of strategic leadership deal with how organizations are led as a whole, i.e. the exercise of indirect leadership over an institution. This body of knowledge is rather inclusive. Subsets of strategic leadership include:

- Vision
- Decision-making
- Organizational processes, structures and control mechanisms

# CHAPTER 4

- Development of successors
- External relations
- Organizational ethics and Culture.[68]

Due in no small part to the origins of the 1980s renaissance in strategic leadership, 'New Leadership' theories, especially with regard to vision, tend to dominate the strategic leadership literature.[69] This is due to the indirect nature of leadership at the organizational level where leaders seldom get the opportunity to communicate face-to-face with all of their subordinates. As a result, a means of passing on direction or conveying intent to their direct subordinates becomes crucial.

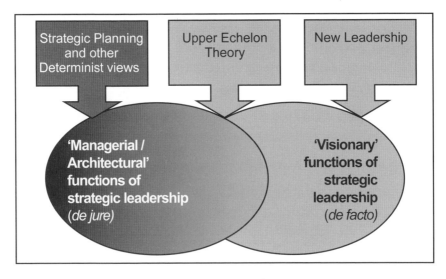

Figure 11: Origins of Strategic Leadership

So what exactly is strategic leadership? It is more than the simple equation of strategy and leadership. Its theoretical basis is that organizations are a reflection of their leadership.[70] This connotes that the adjective 'strategic' refers to the highest level of an organization, meaning the level at which all activities are controlled in a corporation. The use of the term leadership indicates that it is a "relational endeavour",[71] or interaction between two or more people. The loose consensus on the definition of leadership is not entirely helpful here, as strategic leadership also includes management. The line between leadership and

# 4 CHAPTER

management is blurry, and some authors use the terms 'charismatic' to describe the exercise of *de facto* leadership and 'architectural' to describe the exercise of *de jure* leadership instead of the terms of leader or manager.[72] From here on, the term 'Managerial/Architectural' will be used to describe the organizational aspects of strategic leadership in the Business Community.

Regardless of the chosen terminology, the following example illustrates the point about the requirement for 'managerial/architectural' skills:

> Strategic leadership is directing and controlling rational and deliberate action that applies to an organization in its most fundamental sense: purpose, culture, strategy, core competencies, and critical processes. Strategic leadership includes not only operating successfully today but also guiding deep and abiding change – transformation – into the essence of an organization.[73]

There are two subsets to the concept of strategic leadership. The first is the future-oriented, shared intent-driven 'visionary' subset and the second is the 'managerial/architectural' subset that emphasizes the requirement for stability and processes in the present.[74] The 'visionary' subset springs from 'New Leadership' thinking whereas the 'managerial/architectural' subset is a reflection of 'Upper Echelons' theory as well as some of the Environmentally Driven paradigm as depicted in Figure 11 (page 39). The determinist views on strategy place an emphasis on management and controls in order to unify organizations in the face of the challenges posed by the environment.

The 'visionary' subset is more prevalent in the literature than the other. The following examples show the emphasis on the relationship between the leader and the future fortunes of the organization:

> ...multifunctional, involves managing through others, and helps organizations cope with change that seems to be increasing exponentially in today's globalized environment...requires the

> ability to accommodate and integrate both internal and external conditions, and to manage and engage in complex information processing.[75]
>
> If strategy is defined as the patterns of choices made to achieve a sustainable competitive advantage, then strategic leadership involves focusing on the choices that enhance the health and well-being of an organization over the long term.[76]
>
> ...nothing more than the ability to anticipate, prepare, and get positioned for the future. It is the ability to mobilize and focus resources and energy on things that make a difference and will position you for success in the future...[77]
>
> Strategic leadership is defined as a person's ability to anticipate, envision, maintain flexibility, think strategically, and work with others to initiate changes that will create a viable future for the organization.[78]

The pattern here is very clear. Strategic leadership is about a leader's or group of leaders' actions and decisions taken in the present to improve the organization's fortunes in the future. The language suggests that this is a relatively distant thing, but the time horizon has not been defined.

The debate between the Environmentally Driven and the Choice Driven paradigms also manifests itself in the strategic leadership literature. Elements of this literature assign the role of making sense of the environment to the strategic leader. For example, the Center for Management & Organization Effectiveness argues that: "The primary goal of a strategic leader is to gain a better understanding of the business conditions, the environment (the market, customers and competitors), and the leading indicators that identify new trends and situations that may arise..."[79] In another example, it is argued that strategic leadership is fundamentally about planning, where the leader engages in a series of steps:

# 4 CHAPTER

- Understand the environment
- Build a strategy on that understanding
- Plan activities on the strategy
- Ensure resource allocation matches that strategy.[80]

Another example posits that leading an organization is about adapting it to the environment.[81] These examples suggest that strategic leadership is more about strategy than it is about leadership. There is a contradiction here. Leadership is assumed to exist as a *de jure* function yet the literature emphasizes the exercise of *de facto* leadership.

Strategic leadership in the Business Community can be classified in a number of manners. First, its nature is both cognitive, in that it deals in outcomes and how to achieve them (i.e. strategy as a plan), and affective, in that it makes appeals to human values (i.e. *de facto* leadership).[82] Another way of looking at the classification is to treat strategic leadership as being comprised of three types of leadership:

- 'Meta-leadership' – involving the use of vision and institutional stewardship
- 'Micro-leadership' – involving the development and maintenance of human relationships and influence over others
- 'Macro-leadership' – involving the use of strategic goals and how one organizes to meet them.[83]

Strategic leadership would require all three types of leadership described above. This approach to classification is very inclusive. Note that it includes the use of 'New Leadership' tenets (*de facto* leadership) and the requirement for 'managerial/architectural' skill (planning). The existence of *de jure* leadership is assumed. Each element is treated as distinct yet complementary to each other.

The same inclusive approach can be seen in a discussion of 'frames' used by strategic leaders. 'Frames' are a cognitive device to allow individuals, in this case, leaders to focus on the important aspects of something under observation, make a cursory analysis, to help them make decisions.[84] In his commentary on 'frames', Richard Hughes of the Center for Creative Leadership suggested that strategic leaders maintained three separate 'frames' of the world around them. In the first, they maintained a perspective on their own characteristics, be they strengths or weaknesses. In the second, they examined the 'Competitive Environment' by looking at a series of external factors to develop a deeper understanding. Lastly, they maintained an organizational 'frame' to be aware of the strengths, weaknesses and other characteristics of their own organizations.[85] This demonstrated a balance between the choice-driven and environmentally driven paradigms on the issue of strategy. Leadership matters, but only so long as it is capable of steering the organization through the environment.

A strategic leader, according to the literature, must be capable of performing many different tasks. Naturally, the importance of the tasks varies depending on the school of thought. The 'New Leadership'-oriented authors would describe the more important tasks as those that rely on the creation and communication of a vision. One author offered the concept that strategic leaders were really the 'leaders of leaders', and as a result, they needed, above all else, to communicate a vision and create a framework for decentralized decision-making within their organizations.[86] One pair of writers argued that: "Strategic leadership provides the vision, direction, the purpose for growth, and context for the success of the corporation...it provides the umbrella under which businesses devise appropriate strategies and create value."[87] In another example, John Adair, listed a series of tasks, all of which are team-oriented:

- Building and maintaining the team
- Institutional stewardship
- Achieving the common task
- Selection and maintenance of the aim
- Task assignment
- Motivating and developing the individual.[88]

The difference with 'New Science' and 'New Leadership' thinking and other parts of the strategic leadership literature is that it tolerates the risk of unproductive work

# 4 CHAPTER

associated with empowerment. Vision alone is considered to be a sufficient degree of control as opposed to vision and a series of processes and policy controls.

Those with greater 'managerial/architectural' inclinations are far less willing to let the vision be the guide for all efforts. Strategic leadership is about leading an organization, and this means that the exercise of such leadership is an effort to clarify the situation facing the organization and to set the conditions for success in planning and execution of missions. The conditions for success, in this stream, are set through the development, maintenance and enforcement of organizational controls such as policies, regulations, frameworks, or structures.[89] This provides a means of channelling the efforts of entire organizations to the goals and reducing unproductive effort. In such schemes, leadership's role is therefore to pass on its knowledge of strategy, define and communicate the firm's 'unique position', make decisions and force the organization to fit with the demands of the environment.[90] In another example, two strategic leadership experts from the Center for Creative Leadership, Richard Hughes and Katherine Beatty, counsel that a five-step process is sufficient to describe the role of strategic leadership:

- Clarify aspirations and business strategy
- Identify capabilities to implement business strategy
- Assess those capabilities
- Make leadership development a key component
- Get top leadership support.[91]

The important point to note with these last comments was that the authors have observed that part of a strategic leader's role is to generate other strategic leaders and/or coordinate a series of plans. The prescribed passage of knowledge of strategy to one's subordinates is a means of ensuring that the organization continues to be successful. The issue of the development of potential strategic leaders in the Business Community will be addressed later.

There are a number of different approaches that a strategic leader can adopt, and it is inferred that the best strategic leaders use all of them. The approaches were gleaned from a series of different methods and their relative importance depended on the method used to gather the information. The first approach, which used survey data, suggested that the descriptive approach using the concept of a vision was the most effective means. The reporting of introspection by select business leaders led those that engage in the art of strategic leadership to state that the use of language to frame problems and craft solutions was more effective, yet those same individuals, when interviewed, declared that vision was far more effective. Interviews with their subordinates revealed that the subordinates placed an emphasis on how strategic leaders set an example for others to follow.[92] The data is interesting. Strategic leaders, in private, stated that how they use language was more important than the use of a vision, but in surveys and interviews, both strategic leaders and their subordinates attached a greater degree of importance to the use of vision. It is suspected that this is a case of the results being influenced by a perceived need to follow the trend in the literature that emphasizes the utility and relevance of vision. It appears that the Business Community's literature influences how its adherents perceive and consider leadership, but that the practice of leadership is actually more constant and coherent.

**Development of Strategic Leaders**

Having examined the role of strategic leadership and some of the approaches, it would be useful to examine more closely what competencies strategic leaders require. This is a combination of the skill sets and role requirements. In terms of the former, the literature tends to favour 'New Leadership' thinking. It not only emphasizes the importance of 'influence skills', such as maintaining positive relationships within and with other organizations, but also acknowledges the existence of 'bureaucratic politics'.[93] In another example, it argues that it is not their position or title, but rather their level of responsibility, which defines strategic leaders. Commensurate with one's level of responsibility comes the requirement to stimulate rather than control one's subordinates. According to Ireland and Hitt, interactions with the subordinates ought to be based on sharing 'insights, knowledge and responsibilities for achieved outcomes'.[94] This is inspired by empowerment theory where leaders are responsible to set the conditions for subordinates to succeed.

# 4 CHAPTER

The examination of required skill sets leads one to note that there are other influences on the concept of strategic leadership in the Business Community. The two authors from the Centre for Creative Leadership, Hughes and Beatty, suggest that there are really three types of skill sets required by strategic leaders. First, they need to be future-focused. Second, they need to be change-oriented. Third, they need to think systemically, that is, be capable of understanding organizations as a series of systems.[95] This emphasis on systems is a product of 'New Science' thought on strategy and leadership. This example is by no means an isolated phenomenon within the Business Community, or the American military for that matter. The inclusion of multiple schools of thought accounts for the difficulty in understanding the Business Community's use of the concept of strategic leadership.

There has been some discussion of competencies required by strategic leaders in relation to their position and/or role in an organization. Some of this literature runs the risk of confusing strategic leadership with those that occupy senior positions in any commercial organization by attributing virtue to the position. For example, it has been argued that an individual in a senior position is better able to understand the nature of environment, and therefore is able to create and foster a vision, select capable people to populate the organization and create a positive organizational culture.[96] The cause was attributed to the position as opposed to the individual's level of experience, talents or intellect. Such attributions are misleading and fuel observations that strategic leadership is another manifestation of 'great man theory'.

Role-based descriptions are less prone to such criticism, but blur the distinction between management and leadership. For example, one author offered a list of different roles associated with strategic leaders:

- Classical Administrator – The leader acts as an organizer in accordance with the tenets of 'Scientific Management'.
- Design School Planner – The leader acts as a planner in accordance with the tenets of 'Strategic Planning'.

# CHAPTER 4

- Role Playing Manager – The leader plays a series of different roles and adapts to the situation at hand.
- Political Contingency Responder – The leader's choice of roles is based on improving or maintaining power over others.
- Competitive Positioner – The leader observes the environment and responds to its demands.
- Visionary Transformer – The leader seeks to consistently improve the organization.
- Self-Organizing Facilitator – The leader acts as an organizational designer.
- Turnaround Strategist – The leader seeks to improve the organization rapidly.
- Crisis-Avoider Strategist – The leader seeks to ensure that the organization does not fall victim to crises and minimize the duration and damage caused by crises.[97]

Most of these roles are a reflection of the various schools of leadership and strategy. Some, however, are borne of necessity imposed by the surrounding environment.

The 'New Science' influence over strategic leadership can also be seen in some of the literature referring to competencies required by strategic leaders. In order to make sense of the environment around them and their organization, strategic leaders need to be able to multi-task and identify patterns of cause and effect quickly. This has led some to argue that the key competency for strategic leaders is the development and maintenance of cognitive complexity.[98] This provides strategic leaders with the ability to identify and exploit opportunities for their organization. This is an interesting point as it represents an overlap with the American military. This is similar to the concept of the *Coup D'Oeil*, a concept that also refers to a General's ability to understand the battlefield after one glance and make crucial decisions. Other authors have argued that successful strategic leaders have been able to exploit '*Kairos*' time, which refers to the ability to take the right action at a critical time.[99] The two are very similar concepts.

# CHAPTER 4

**Is Strategic Leadership a group or individual function?**

Most of the literature addresses strategic leadership as an individual act as opposed to a group function. The same point can be made about the criticism levelled at strategic leadership. Both of these observations raise the question of whether or not the Business Community sees strategic leadership as a group function or individual act. This is an issue based on two different views of how strategic leaders operate. On the one hand, an omnipotent CEO is alone to lead an organization in a predictable environment and on the other hand, the CEO and his team are responsible to lead an organization in a tumultuous world.[100] The first view tends towards the Choice Driven paradigm and the second towards the Environmentally Driven paradigm.

Very little of the literature directly prescribes that strategic leadership is an individual act. It can, however, be argued that most of the literature makes this implication. The critics of the concepts of strategic leadership make it rather explicit and base the majority of the arguments on the individual nature of strategic leadership, i.e. a form of hero worship. For example, Margaret Wheatley, one of the advocates of 'New Science', argues that the Newtonian view of the lonely universe has contributed to the 'great man' or heroic theory of leadership, and 'New Science' uses a more sophisticated view of the universe, therefore, is less prone to adhering to the 'great man' theory of leadership.[101] This theory of leadership tends to transfer responsibility to individual leaders for everything that occurs within an organization. Strategic management has a bad habit of attributing the causes of success or failure largely to the executives.[102] This reduces the cause to a single factor as opposed to a range of factors. 'Upper Echelon' theory can be seen in this light. According to this theory, organizational outcomes are dependent on senior management choices, which would make it a theory of decision-making. The evidence to support 'Upper Echelon Theory' was based on demographic data to attribute causality to their decision calculus as opposed to actual psychological profiles.[103] This suggests that one of the major sources of strategic leadership thought in the Business Community is based on a theory propped up with rather subjective evidence.

# CHAPTER 4

Proponents of the notion that strategic leadership is a group function have focused on three aspects associated with 'New Leadership'. These aspects are: transformational leadership, cultural biases and compromises. 'New Leadership' is related to 'New Science' in that it offers ways for organizations to deal with a complex environment. Empowerment theory is one example, and transformational leadership provides a clearer example. Vision and clear communication of organizational goals provide a means for an organization to harness all the efforts and capabilities of all levels.[104] The point about cultural bias may be a reaction to the rise of European and Japanese organizations. Those seeking explanations for the rise of corporations from Europe and Japan while American and British enterprise seemed to decline argued that the Anglo-American tradition of 'Great Man' theory could not compete with the team-oriented Japanese or European approaches.[105] This attributes virtue to the group approach while tying the 'Great Man' theory to failing organizations. Some, such as the Center for Creative Leadership, adopted the compromise of arguing that it was both an individual act and a group function.[106] The issue of whether or not strategic leadership is a group function or individual act is itself plagued by the lack of common definitions. The authors of a study on 'Upper Echelon Theory' noted that the definition of 'Top Management Teams' in the literature varied significantly.[107] It would appear that the arguments stating that strategic leadership is a group function are also limited by the weakness of the supporting evidence. It must be noted that the weakness of the evidence stems from the imprecise use of language.

The discussion of the debate on strategic leadership as a group or individual act in the previous paragraphs revealed that in the Business Community, strategic leadership is based on theories. These are based on questionable assumptions and imprecise language and their resemblance to reality may be declining despite their popularity.

**Generation of Strategic Leaders**

The Business Community is much weaker than the American military in addressing the issue of how to develop and maintain strategic leaders within

# 4 CHAPTER

organizations. It was remarkable to find numerous articles offered by strategic leadership experts on how to be a strategic leader, but how few there were on how to teach others to be strategic leaders. There are four examples of the latter that merit mention. The first example was incredibly simple; it prescribed that an organizational structure ought to be matched with people over time through individual evaluation and development.[108] The second example offered that the development of strategic leaders was based on mentoring of subordinates to allow their strategic thinking and core competencies to flourish while also setting the conditions in the organization to encourage strategic leadership, such as rewards, controls and a supportive culture.[109] The third was similar in that it prescribes both individual and collective talent pools should be developed as a means to compete with organizations.[110] The final example was more detailed in that it prescribed six experiences for potential strategic leaders:

- 360-degree feedback
- Feedback intensive programs
- Skills based training
- Challenging job assignments
- Developmental relationships
- Hardship.[111]

This is a more detailed list and it bears more than a passing resemblance to some of the Five American Armed Services' professional development frameworks, but it is a relatively rare thing to see.

## Part 4: Summary

With regard to strategic leadership, the Business Community is like a swirling vortex, and it is very possible for one to become lost in its depths. There is no set definition of the term of strategy and efforts to define it can draw one quickly into lengthy, if not interminable, debates over its nature. The loose consensus on leadership being an influence process (i.e. *de facto* leadership) provides some relief, but this is short-lived. When one begins to combine the two to arrive at an understanding of strategic leadership, one finds that the debates over strategy

# CHAPTER 4

permeate the strategic leadership literature and that the study of strategic leadership developed from a paradigm of leadership with a weak theoretical foundation. To the Business Community, strategic leadership has less to do with the concept of strategy as it does with ideas about leadership of an institution.

# Chapter 5

# *Scylla*: The Five American Armed Services

In this monograph, the American military services have been cast as Scylla the five-headed terror.[1] It could have been cast as Charybdis except for one thing: at least four of the Five American Armed Services (Army, Navy, Air Force, Marines and the Coast Guard) share a common definition of the term 'strategy'. This shared definition is like Scylla's body but each of the heads (the Armed Services) has a different view of what leadership ought to be, service ethos and how leadership is developed in their service.

## Part 1: Strategy in the Five American Armed Services

### Strategy in Politico-Military Context

The Five American Armed Services are different from the Business Community in that the definition of the term strategy and the collective understanding of the concept are very clear, as the Joint Chiefs of Staff have defined them. Due to the nature of the American polity and American civil-military relations, the Armed Services have a set role in the United States. This role has been defined explicitly in American military doctrine. Joint Publication 1 (JP 1), *Joint Warfare of the Armed Forces of the United States*, frames the concept of strategy for all of the Armed Services. It describes the military's role in strategy formulation and implementation. The National Security Strategy (NSS) is a policy statement of how the American government will ensure the security of the United States and its citizenry. It outlines the government's national security objectives and how the government will use all of its diplomatic, economic, informational and military instruments of power in pursuit of those objectives. The National Military Strategy (NMS) describes how American military assets and resources will be employed to

# CHAPTER 5

achieve the national security objectives laid out in the NSS.[2] The NSS represents a statement of American policy and the NMS is a subordinate document describing American strategy. The NMS is the highest-level military document in the United States; yet it does not enter the political realm but translates policy into military direction and resource allocation. The strategic level of war, for the American military, provides the 'bridging' function between policy and military affairs.[3] It should be noted that the 'bridging' of policy and military operations is not described in the OCD, but it has been captured in the academic literature related to military affairs.

Not surprisingly, this 'bridging function' of strategy is a partial product of the American historical experience in war and its effect on American strategic thought. Antulio Echevarria of the U.S. Army War College's Strategic Studies Institute argued that there is:

> ...a persistent bifurcation in American strategic thinking – though by no means unique to Americans – in which military professionals concentrate on winning battles and campaigns, while policymakers focus on the diplomatic struggles that precede and influence, or are influenced by, the actual fighting. This bifurcation is partly a matter of preference and partly a by-product of the American tradition of subordinating military command to civilian leadership, which creates two separate spheres of responsibility, one for diplomacy and one for combat.[4]

In short, the American strategic tradition conceives of states of war or peace; that is war is either fought on a grand scale by the military or political issues are resolved peacefully through the exercise of diplomacy. The use of force and diplomacy do not co-exist in this view, which creates the conditions for the NSS/NMS policy construct. The division between the NSS and NMS is based on the total war construct, and suggests that the 'bridging function' represents an ideal state. The reality of the strategic level of war is not so clear.

# CHAPTER 5

The American military establishment's understanding of strategy is heavily informed by the relationship between the NSS and the NMS. The NSS sets the policy goals and assigns them to elements or wings of the American government. The goals are an output of the NSS and represent an input for the NMS as depicted in Figure 12 below. Policy goals, in the American military vernacular, are described as 'Ends'. The NMS takes the assigned 'Ends', translates them into a series of military objectives, and assigns 'Ways' (or courses of action to achieve an objective) as well as the 'Means' (or resources to achieve an objective).[5] Strategy, for the American military establishment, is all about the association of ways and means to achieve the ends. Yet this is far too tidy. War is a dynamic process and enemy strategies and actions may interrupt the simple interaction between ends, ways, and means. Clausewitz's writings on the subject have been paraphrased as "...the *purpose* of war is to serve policy, but the *nature* of war is to serve itself."[6] The heuristic device of ends, ways and means addresses the definition well to the first, third and fourth OCD definitions (i.e. strategy as a long range plan, as a stratagem or the art of planning and directing military activity).

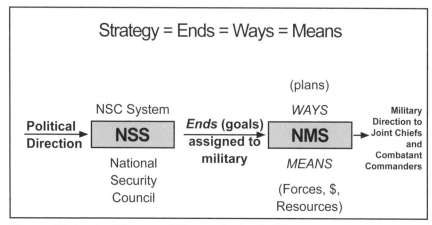

Figure 12: Ends, Ways and Means in the NMS

**Strategy and the Five American Armed Services**

Even with a joint definition, the concept of strategy has different nuances in each of the Armed Services. These nuances are born of different requirements and

# 5 CHAPTER

associated organizational cultures. Each definition reveals different things about each of the services and their concept of strategy.

The U.S. Army does not distinguish between grand strategy and strategy. Its capstone publication of doctrine, Field Manual (FM) 3.0 Operations, states that: "***Strategy* is the art and science of developing and employing armed forces and other instruments of national power in a synchronized fashion to secure national or multinational objectives.**"[7] This is close to the first, third and fourth OCD definitions (i.e. strategy as a long range plan, as a stratagem or the art of planning and directing military activity) and is close to the original Ancient Greek definitions. This suggests that the U.S. Army does not wish to become embroiled in the higher levels of war as that is the province of the Chairman of Joint Chiefs of Staff and the Combatant Commanders, but focuses its doctrine at the tactical and operational levels.[8] There is also an institutional dimension. The Army is responsible, as a Title 10 function, to organize, equip and train its forces.[9] As a result, it draws a distinction between the 'Institutional Army' and the 'Operational Army'. The former is comprised of the infrastructure and capabilities that allow for the elements of the latter to be organized, equipped and trained. There is a dichotomy at work here; FM 3.0 prescribes strategy for the 'Operational Army' (therefore fulfilling the fourth OCD definition of strategy as the art of planning and directing military activity) but the Title 10 responsibilities orient the 'Institutional Army' on a different form of strategy (based on the first OCD definition of strategy as a long-range plan).

The U.S. Air Force's definition is similar to the U.S. Army's in that it does not make any significant distinction between the policy and strategic levels. Its language is simple and to the point: "Strategy defines how operations will be conducted to accomplish national policy objectives".[10] Strategy is about the ends/ways/means relationship. It relates how the ends (national policy goals) will be achieved through an association of ways and means (the courses of action associated with resources known as 'operations'). The U.S. Air Force recognizes all three levels of war and has a simple way to describe them as shown in Figure 13. Strategy is an exercise in matching ends with means and operations and tactics provide the ways. What is different here is that unlike the Army, the institutional focus is at the operational level of war. Due to the nature of airpower, airmen have organized for battle primarily at the theatre level as it affords the best economies of scale for air assets. This geographic delineation has been associated more with the operational level of war. The U.S. Air Force has not made a distinction between its operational and institutional subsets like the Army.

> **LEVELS OF WAR:**
> **AN AIRMAN'S VIEW**
> **Strategic** .................................. **Why, With What**
> **Operational** ............................ **What to Attack**
> **Order of Attack**
> **Duration**
> **Tactical** ........................................................ **How**

Figure 13: AFDD 2 Description of the Levels of War[11]

The U.S. Navy's doctrine does not contain much more than the joint doctrine and uses the concept more as a heuristic device. This may be because the construct of the levels of war was developed for land warfare. The Navy's capstone publication, Naval Doctrine Publication 1, states that:

> Fundamentally, all military forces exist to prepare for and, if necessary, to fight and win wars. To carry out our naval roles, we must be ready to conduct prompt and sustained combat operations – to fight and win at sea, on land, and in the air. Defending the United States and controlling its seaward approaches are the first requirements. Gaining and maintaining control of the sea and establishing our forward sea lines of communication are our next priorities. As we operate in littoral areas of the world on a continuing basis, naval forces provide military power for projection against tactical, operational, and strategic targets. In both peace and war, we frequently carry out

# 5 CHAPTER

our roles through campaigns. A campaign, although often used only in the context of war, is a progression of related military operations aimed at attaining common objectives. Campaigns focus on the operational level of war.[12]

Due to the nature of its employment where ships must be deployed forward from bases to the operating areas, this service also focuses its efforts at the operational level. The concept of the levels of war, to the U.S. Navy, helps demonstrate how naval actions contribute to political outcomes. For example, NDP 1 states that:

> The concept of "levels of war" can help us visualize the relative contribution of military objectives toward achieving overall national goals and offer us a way to place in perspective the causes and effects of our specific objectives, planning, and actions. There are three levels: tactical, operational, and strategic – each increasingly broader in scope. Although the levels do not have precise boundaries, in general we can say that the tactical level involves the details of individual engagements; the operational level concerns forces collectively in a theatre; and the strategic level focuses on supporting national goals."[13]

Note that this service also leaves the strategic realm to others. This is similar to the U.S. Army, but there is no mention in the doctrine about an institutional or an operational Navy.

The U.S. Marine Corps' definition of strategy is framed by the contents of NDP 1. Its definition is very similar to the Army's, but this is more a reflection of the nature of land warfare. Marine Corps Doctrine Publication 1 (MCDP 1) describes strategy as the 'way' to bring ends and means together, and it is confined to the realm of military strategy.[14] MCDP 1 provides additional details on the concept and seeks to show relationships between the political ends and military ways and means. It argues that when military means are unlimited, a stratagem of annihilation will be pursued. Such stratagems seek to:

## CHAPTER 5

...deprive the enemy of the ability to resist, to make him militarily helpless. Annihilation does not require the complete physical destruction of the enemy's military forces. Rather, it requires that the forces be so demoralized and disorganized that they become unable to effectively interfere with the achievement of our political goals. What is being annihilated – literally "made into nothing"– is the enemy's physical means to oppose us.[15]

On the other hand, when military objectives are limited, a stratagem of erosion will be pursued. This stratagem seeks: "to convince the enemy that settling the political dispute will be easier and the outcome more attractive than continued conflict. To put it another way, erosion strategies seek to present the enemy with the probability of an outcome worse *in his eyes* than peace on the adversary's terms."[16] The inclusion of the term 'stratagem' is useful as it illustrates a broader point – even the Five American Armed Services uses the term 'strategy' to cover both 'strategy' and 'stratagems'.

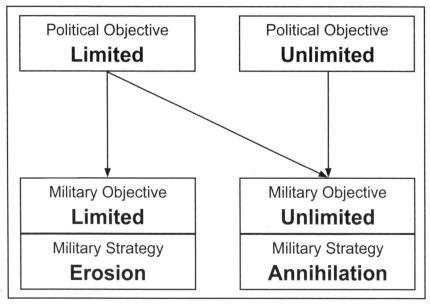

Figure 14: Political and Military Objectives and Stratagems[17]

# 5 CHAPTER

The U.S. Coast Guard is the one exception to the rule about the Five American Armed Services. It does not recognize strategy in the same manner as the other four services. Its activities contribute to the achievement of the NSS, but it does not view its activities through the lens of the levels of war. The Coast Guard does not have to do so due to its special legal status. It operates primarily as a law enforcement agency but falls under the U.S. Navy if war is declared or if the President directs.[18] It does, however, associate the concept of strategy with the institution.

**Strategy for the Institution**

The Five American Armed Services live within the confines of a democracy. It can be argued that their preference to focus on the operational level is driven by the need for military institutions within democratic states to be somewhat 'astrategic', i.e. they focus on the preservation or enhancement of their force structure and the government, who are open, impatient and appear to be lacking coordination, need to maintain oversight on military activities. This creates a tension between the military requirement to preserve and improve the institution and the political requirement to ensure the institution does not consume more than its share of the resources and stays in line with the body politic.[19] Furthermore, strategy tends towards rationality, meaning an ordered application of efforts based on a coherent list of priorities and democratic states, by their very nature, must frequently change, or at least be seen to change, efforts and priorities in response to the demands of their electorates and therefore, appear to be irrational.[20] The institutional dimension of strategy becomes far more important as result. Furthermore, the Armed Services have based their doctrine on a framework of the levels of war. The divisions between the levels, however, are not very clean and this can blur discussions of leadership.[21]

**Dimensions of Strategy**

The Five American Armed Services also view the concept of strategy in terms of different time horizons and scale. The strategic level comes with significant responsibilities, and those working at that level must provide for national defence with the rest of the government apparatus and with American society, to set the conditions for future capability and they must also manage combined (i.e. other militaries) and joint (i.e. other services within the American military) relationships.[22] These responsibilities, it should be noted, are all associated with the institution as opposed to the strategic level of war. The literature, like that in the Business Community, also suggested that decisions at the strategic level had longer time horizons and a greater magnitude.[23] This comment applies equally to the institutional dimension of strategy as the strategic level of war.

The Five American Armed Services differ from the Business Community is that the debate between the two paradigms (Choice Driven versus Environmentally Determined) does not exist. Truth be told, four of the five services do not consider the debate in any significant manner. The exception to the rule is the U.S. Army, and it is clear from its body of doctrine that the Environmentally Determined paradigm reigns supreme. The environment, for the U.S. Army, can be sub-divided into a series of specific environments as follows:

- National Security
- Domestic
- Military
- International

Each of these environments should be considered 'open systems' as events in one environment affect them all in surprising ways. The environment is therefore volatile, unpredictable, chaotic and ambiguous (VUCA).[24] This view of the environment has roots in 'New Science' and its applications to war. Some authors argue that war, as it known today, ought to be fought by military organizations using shared but dispersed information acting as complex adaptive systems. The results of such conflicts are difficult to predict in advance and are non-linear.[25] The existence of the 'Contemporary Operating Environment' in which the current threat is primarily asymmetric, e.g. a terrorist or insurgent reinforces this notion. For the Five American Armed Services, whose heritage was a monolithic Soviet-led conventional force, this 'Contemporary Operating Environment' appears to be even more VUCA.[26]

# CHAPTER 5

To imply that there is no debate about VUCA within the American military's ranks would be misleading. One can find arguments and criticisms of the concept, but these do not come from the perspective of the Choice Driven paradigm. There are two sources of criticism: one, some critics point out that the body of doctrine associated with VUCA is incoherent, and two, the ramifications of VUCA at the strategic level are counter-intuitive to professional officers.

VUCA describes the nature of the environment as being chaotic. The remainder of Army doctrine is based on the notion that military planning, whether it is at the strategic, operational or tactical levels, is a deliberate and rational process intended to develop an ordered approach.[27] Another author pointed out that the doctrine, despite the VUCA nature of the environment, seeks to maintain a centralized system with a very strong emphasis on structure.[28] If the environment is VUCA, is the doctrine seeking to address or ignore the issue? These criticisms imply that the U.S. Army, and by inference, the other four Armed Services, have not transformed themselves into 'complex adaptive systems' in sufficient measure. It would be unrealistic, however, to conceive of a military institution to become a 'self-organizing system' the same way a business organization could. Structure and hierarchy are crucial to a military organization and by nature; they are conservative and tend to resist change.

The second criticism is pointed more towards the development of strategic leaders than the Five American Armed Services' concept of strategy, but the concept of strategy and the development of future strategic leaders cannot be separated easily. VUCA, like any environmentally determinist concept, is based on the idea that the environment forces adaptation upon military institutions, i.e. an internal response to something that cannot be shaped or altered. To adapt successfully, within an institution, there is a requirement for both consensus and compromise at the strategic level.[29] Neither of these are hallmarks of leadership at the tactical and operational levels. If one takes the institutional view of strategy, VUCA means that strategy's connotation as a rational calculus of ends, ways and means is a heuristic device at best. Given that the ends-ways-means equations can lead to a number of different combinations of means to achieve a defined end, these choices of means will often compete for the same ways. Strategic

# CHAPTER 5

leaders, therefore, must be prepared to compete in the political-bureaucratic battles that surround choices of particular ways, means or both.[30] This is an uncomfortable point for strategic leaders within the United States as it means fighting the battles within the Beltway as opposed to traditional military service. This is culturally driven. Military personnel revere the commanders in the field and can be distrustful of senior commanders and staff officers embedded with large bureaucracies like the Pentagon.[31] The same lack of comfort can be felt in the Canadian Forces with the popular disdain for those senior leaders deemed to have become 'political'.

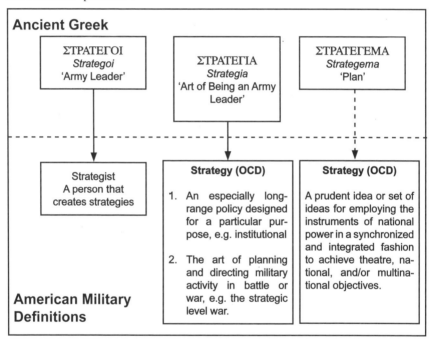

Figure 15: Definition of Strategy in the Five American Armed Services[32]

The Five American Armed Services have a shared definition of strategy, and when compared to the Business Community, it appears to be clear and concise. Strategy has three overlapping definitions within the American military. It is seen as a mental construct combining political ends with military ways and means at the policy level. This definition matches the official Department of Defense definition

# 5 CHAPTER

of strategy shown in Figure 15 (page 63). Two of the other definitions bear some resemblance to the OCD definitions, yet there are still connotations to each of these. While the Five American Armed Services view strategy as a long-range plan, this definition is associated with the institution of that armed service as opposed to strategy as a level of war. The last of the definition is closer to the definition of strategy as planning and directing military activity at the strategic level of war.

## Part 2: Leadership in the Five American Armed Services

Analyzing military leadership doctrine is a complex task, as it is difficult to discern between the various concepts of command, leadership and management. Understanding the nature of and the relationship between these three concepts is vital to comprehending how a military institution views and applies these concepts. The Business Community, for example, appears to hold that leadership is about influencing others whereas management is about organizing others. The American military also has to contend with concept of command, which may mean that legal structures that developed in order to ensure the compliance military personnel with their superior's wishes complicate the issue. The joint military dictionary states that the first definition of command is:

> (DOD) 1. The authority that a commander in the Armed Forces lawfully exercises over subordinates by virtue of rank or assignment. Command includes the authority and responsibility for effectively using available resources and for planning the employment of, organizing, directing, coordinating, and controlling military forces for the accomplishment of assigned missions. It also includes responsibility for health, welfare, morale, and discipline of assigned personnel.[33]

Command is a legal relationship between the commander and the commanded where the commander is responsible for both tasks one might associate with management (e.g. planning, organizing, coordinating, controlling) and leadership (e.g. responsibility for health, welfare, morale and discipline). It is the ultimate in the *de jure* dimension of leadership. It is possible to make the argument based

# CHAPTER 5

on those definitions of command, leadership and management that the American military has the solution to the Business Community's problem with the dichotomy between leadership and the management/'architectural' roles associated with strategic leaders. To make matters worse, the American military lexicon, at the joint level, recognizes the concept of command but appears to leave the other two concepts to the services.

**The Five American Armed Services on Leadership**

The U.S. Army's doctrine on leadership is encapsulated in FM 22-100 *Army Leadership: Be, Know, Do*. This field manual states that: "Leadership is **influencing** people – by providing purpose, direction, and motivation – while **operating** to accomplish the mission and **improving** the organization."[34] This definition differs somewhat from the joint definition of command. It is consistent with the Business Community with regard to leadership as an influence task; this is at odds with the view that organizational tasks (e.g. direction and improvement) are related to command. The definition also takes into account that leaders must often balance the requirements of the mission and the needs of their subordinates. The last sentence fragment also suggests that U.S. Army leaders have a duty to improve their organizations, which implies both seeing to the needs of their subordinates and the long-term health of the organization. This reinforces the point that the Five American Armed Services tend to favour the institutional definition of strategy when dealing with leadership.

The U.S. Army's view has aroused some criticism. The inclusion of managerial concepts within FM 22-100 has led some to accuse the doctrine of confusing *de jure* with *de facto* leadership.[35] This means that technical skills play a greater role within the construct of *de facto* leadership. This criticism, however, is based on the notion that followers in any organization can withhold their allegiance easily or at a whim. While such things can occur, the authority relationships inherent within the *de jure* dimension of leadership hold them in relative check, and the end result may be begrudging compliance in the case of poor leaders.

# 5 CHAPTER

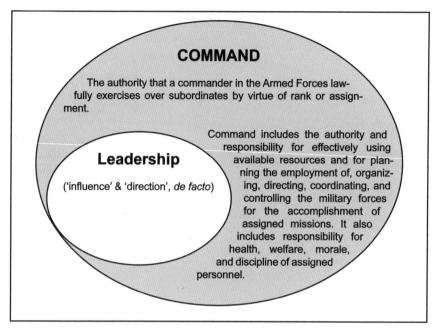

Figure 16: Command, Leadership and Management[36]

The U.S. Air Force's view of leadership differs somewhat from the Army's view. It must be noted that the Air Force's culture has been shaped by the nature of its operations where the fighting is carried out by aircrew, who are primarily officers, while all of the others are relegated to support roles.[37] Leadership, for the U.S. Air Force, is: "...the art and science of influencing and directing people to accomplish the assigned mission."[38] Like the Army, it appears to blend managerial/'architectural' and influence tasks and is focused on the accomplishment of assigned missions. The U.S. Air Force's key leadership manual argues that leaders must balance between the use of personal, team and institutional leadership, and at the strategic level, the emphasis is on the latter.[39] The use of a particular type of leadership is prescribed in order: "...to establish structure, allocate resources, and articulate strategic vision."[40] Two of the three of these tasks are managerial/'architectural' in nature as they address organizational issues. What this suggests is that this service views strategic leadership differently than at other levels.

# CHAPTER 5

| Collection | Leadership | Management and Strategic Planning |
|---|---|---|
| Junior Enlisted | - Donald Phillips, *Lincoln on Leadership*<br>- Robert Heinlein, *Starship Troopers* | - Stephen Covey, *7 Habits of Highly Effective People*<br>- Julie Morgenstern, *Time Management from the Inside Out* |
| Leading Petty Officer | - Margot Morrell and Stephanie Capparell, *Shackleton's Way: Leadership Lessons from the Great Antarctic Explorer*<br>- Herman Melville, *Billy Budd and Other Stories* | - Malcolm Gladwell, *The Tipping Point: How Little Things Can Make a Big Difference*<br>- Robert A. Heineman, *American Government* |
| Division Leaders | - James S. Hirsch, *Two Souls Indivisible*<br>- C.S. Forester, *The Good Shepherd* | - Steven D. Leavitt, *Freakonomics: A Rogue Economist Studies the Hidden Side of Everything*<br>- Clayton M. Christensen, *The Innovator's Dilemma* |
| Department / Command Leader | - Patrick O'Brian, *Master and Commander*<br>- Barry McCaffrey, *The Warrior's Art* | - Marilyn Loden, *Implementing Diversity*<br>- Larry Bossidy, *Execution: The Discipline of Getting Things Done* |
| Senior Leaders | - Rudolph W. Giuliani, *Leadership*<br>- Winston Churchill, *The Second World War Volume 1: The Gathering Storm* | - Michael Lewis, *Moneyball: The Art of Winning an Unfair Game*<br>- Henry Mintzberg, *The Rise and Fall of Strategic Planning* |

Table 3: Navy Reading List on Leadership, Management and Strategic Planning[41]

# 5 CHAPTER

More research appears to have been done into the evolution of the Air Force's leadership doctrine. This evolution closely matches the evolution of prevailing leadership theories as well as the type of war that was seen as likely during any given timeframe. U.S. Air Force leadership doctrine has been shaped by its heritage. The early work on Air Force leadership was designed to support its organizational independence

The Air Force sought to have a psychological justification for its leadership practices in the late 1940s. As the Air Force's role became fixated on strategic deterrence in the following decade, the leadership doctrine was based on theories of group behaviour. The tactical nature (at least in terms of the employment of air power) of the Vietnam War eroded the doctrinal foundation for group behaviour based leadership and like the Business Community, the Air Force based its leadership on systems approaches. Leadership doctrine became decentralized by the 1970s, and it was not until the mid-1980s that it began to be realigned in step with the development of the Air-Land Battle doctrine.[42] In short, there is a greater degree of alignment between fighting doctrine and leadership doctrine.

The U.S. Navy, of all the services, has the fewest publications pertaining to leadership. While in recent years, the Chief of Naval Operations has directed that the U.S. Navy improve its leader development and human resource practices to remain competitive, gaps in its doctrine still exist. One serving naval officer commented in a staff college paper that: "Unlike her sister services, the Navy lacks formalized doctrine concerning virtually any subject with the exception of Naval operations and tactics."[43] This relative paucity of publications makes it difficult to provide a cogent analysis of Navy leadership doctrine. Other sources, such as the Navy's professional reading program offer limited insights into the Navy's view of leadership. See Table 3 (page 67) below for the contents. Note that the collection makes reference to elements of literature from the Business Community. Mintzberg's work, for example, is a fundamental rejection of strategic planning as a concept, favouring instead the development of an emergent strategy as a means to deal with the challenges posed by the environment. It is suspected that Mintzberg's work has been included as a counterpoint to traditional naval planning, i.e. heavily influenced by the long-range and detailed planning associated with shipbuilding.

The Coast Guard's definition of leadership is clear and has been carefully considered. Its definition spans from the tactical to the strategic level and is related to the concept of developing leaders over time. It holds that: "Leadership is the ability to influence others to obtain their obedience, respect, confidence, and loyal cooperation. Good leadership develops through a never-ending process of self-study, education, training, experience, observation, and emulation."[44] Like the Business Community, the Coast Guard leadership assumes that leadership is an influence process. It is also subdivided into four components as follows:

- Self
- Others
- Performance and change
- 'Leading the Coast Guard'

These components are not far off from the concepts of direct and indirect leadership, but it is important to note that they have included the concept of leading the Coast Guard as an institution. This point is stressed in Coast Guard documents:

> As leaders gain experience in the Coast Guard they must understand how it fits into a broader structure of department, branch, government, and the nation as a whole. At a local level, leaders often develop partnerships with public and private sector organizations in order to accomplish the mission. The Coast Guard "plugs in" via its key systems: money, people, and technology. A leader must thoroughly understand these systems and how they interact with similar systems outside the Coast Guard. An awareness of the Coast Guard's value to the nation, and promoting that using a deep understanding of the political system in which we operate becomes more important as one gets more senior. Leaders must develop coalitions and partnerships with allies inside and outside the Coast Guard.[45]

# CHAPTER 5

The only criticism that can be made about the document is that it confuses leadership and management throughout the document.[46] This, however, matches the other services' approaches to command and leadership.

The Five Armed Services showed similar patterns in their definitions of leadership. First, they all sought to inculcate their leaders in the doctrine and ethos of their service. Second, they have all shown some overlap between concepts of leadership and management. Critics of the latter, however, may have forgotten that command at all levels requires a penchant for organization, as these are entities that are intended to fight. Fighting requires that two or more entities collide in a violent and kinetic manner and that a well-organized force often out performs its opponent by virtue of being able to remain organized or to adapt to the conditions. Confusion of leadership and management, in this light, is a virtue as opposed to a vice for a military organization.

## Part 3: Strategic Leadership in the Five American Armed Services

**Concept & Definition**

The Environmentally Driven Paradigm dominates that Five American Armed Services' concepts of strategic leadership. The origins of this difference can be traced to events in military affairs since 2001 and to the perceived rise of asymmetric warfare. Al Qaeda's attack on the World Trade Centre and on the Pentagon was a shock to all and signalled a renaissance of terrorism and insurgency. This, in turn, led to a rebirth of new models of how western states, particularly the United States, ought to fight wars. Various military communities advocated different ways to fight, from the 'Afghan model' which prescribed that airpower and special operations forces be used to support local factions in a war-by-proxy to a traditional counter-insurgency model that prescribed the domination of terrain by hordes of infantry.[47] Such comments would have been unthinkable in the early 1990s. VUCA, as a concept, originated during that time and appears to be increasingly relevant.

# CHAPTER 5

This relevance is reflected in the doctrine manuals, and it is inherently tied to the concept of strategic leadership. The concepts of strategic leadership, however, vary due to the descriptor 'strategic'. Some publications use the concept of strategic as a level of war and discuss strategy solely in terms of the NSS and NMS, whereas others, from the same Armed Service, use 'strategic' in the institutional sense. This suggests that the concept of strategic leadership is less reliant on the definition of strategic than it is on the idea of leaders having to come to grips with a VUCA environment. For example, FM 22-100 contains a chapter dedicated to strategic leadership. It focused on the skills and actions required of a strategic leader in order to be capable of dealing with enacting the NMS in a VUCA environment.[48] This idea of VUCA permeates American military doctrine on strategic leadership. In an article in a USAF journal, Colonel W. Michael Guillot, USAF, makes the explicit argument that strategic leadership is all about planning and executing in the VUCA environment.[49] The same author, however, appeared to try to balance the Environmentally Driven paradigm-driven approach with some 'New Leadership' thought by arguing that while strategic leaders had to deal with the environment, they also need to inspire their subordinates, lead change, engage in critical self-analysis, foster creativity, build teams and maintain a broad consensus.[50] To summarize, strategic leaders need to prepare the internal environment for the demands of VUCA. What may seem like a compromise is actually a prescription for sound preparations.

Some of the Five American Armed Services' views of strategic leadership extend from the definition of strategy as a level of war and seek to elaborate upon the idea that strategy is an alignment between ends, ways and means. This, of course, keeps strategy limited to its national military dimension. In the 1990s, MG Richard Chilcoat, US Army, published his work on 'strategic art' around the same time that the operational art was being discussed within military and related academic circles. MG Chilcoat defined Strategic Art as: "The skilful formulation, coordination, and application of ends (objectives), ways (courses of action), and means (supporting resources) to promote and defend the national interests."[51] This definition places the concept of strategy squarely within the frame of ends, ways and means at the national military level. He stated that: "Strategic Leadership is the effective practice of the strategic art. Strategists can think about and help

# CHAPTER 5

devise strategies, but it is the strategic leader who practices the art and makes it happen."[52] This separates the practitioners from the proverbial 'armchair generals' and would be appealing to many within military circles for that reason alone. One could argue, however, that this confuses leadership with command, as the exercise of 'strategic art' requires *de jure* as well as *de facto* leadership. The three action verbs within the definition cannot occur without the legal sanctioning conveyed with the responsibility of command. His subdivision of strategic art provides further evidence to this effect. There are three overlapping roles within strategic art and true masters of the art act in all three capacities:

- Strategic Theorist – this role is associated with indirect leadership through ideas, teaching or mentoring. Examples include Thucydides, Sun Tzu, and Clausewitz.
- Strategic Practitioner – this role is associated with the planning and execution of strategic activities as a commander. Examples include Patton, Rommel, and Ridgeway.
- Strategic Leader – this role is associated with the provision of vision and focus as well as the motivation of others. Examples include Marshall, Eisenhower, and Churchill.[53]

The examples show a stark division between commanders at the operational level and leaders at the strategic level. It should be noted that one of the examples of a strategic leader was a politician and the other two, while military leaders, by virtue of their position, were thrust into politico-military roles of significant influence. This calls into question some of the American military doctrine on the concept of strategy. It may be the case that the doctrine is normative and described the optimal situation where the military remains within a framework of civilian control unless specific conditions exist that lead political authorities to grant specific individuals within the military a greater degree of latitude.

Not surprisingly, American military literature indicates that the exercise of strategic leadership is dependent on the existence of a hierarchy. It draws a distinction between strategic and lower-level leadership (described as 'direct' or 'general') by noting that strategic leadership is different in terms of 'complexity, time horizon, and focus'.[54] 'Direct' leaders focus on the here and now, and therefore have a very short time horizon, a small span of control over an internal audience and their goals are clear and simple. 'General' leaders have to consider a time horizon of a year to five years, a wider span of control, and both the internal and external audiences while dealing with goals that may be unclear and impossible to achieve. Strategic leaders have to think beyond the twenty-year horizon, wield influence both within and outside the organization and have the necessary conceptual skills to deal with complexity, namely 'systems and integrative thinking'.[55] 'New Science' rears its head as the solution to the challenge posed by VUCA.

Much of the American military thought on strategic leadership can be traced to a single theory developed in the late 1980s. It is related to the 'New Science' paradigm in that it offers a solution to the problem posed by complexity as well as the military requirement to maintain a hierarchical organization in order to prepare to withstand the effects of combat on an organization. This theory, known as 'Stratified Systems Theory' (SST), arises from a general theory of bureaucracy where the complexity of organizational breakdown structures is measured by different time horizons described in the previous paragraph.[56] Like much of the thought in the Business Community, it rests on the Darwinian notion of competition; SST has been described as being: "...primarily a theory of organizational structure in relation to the competitiveness required for survival in a world environment ...a theory of managerial performance requirements derived from that structure and of managerial capabilities necessary to deal with the performance requirements."[57] The theory, it notes, holds that leaders need different skills and skill levels based on where they are employed within an organization. Capabilities are derived from performance requirements, which are derived from the position an individual holds. SST is based on the idea that a hierarchy offers best way to organize to ensure control and accountability within an organization. It also argues that higher echelons of any hierarchy have 'frames of reference' that are more sophisticated and externally focused.[58] While the argument is partially correct in that higher echelons tend to have a greater external focus, it does not explain why these same echelons would have more sophisticated frames of reference. It could be inferred that this was because military institutions operate on a principle of seniority and therefore the higher echelons tend to be populated with extremely experienced personnel. Experience would inform intuition and expand one's frame of reference. Yet the key variable in SST is 'conceptual capacity'.[59] This is somewhat unrelated to experience unless one draws a link between 'conceptual capacity' and frames of reference.

SST, however popular it may be, is not a panacea. It runs the risk of seeing the concept of strategic leaders reduced to being nothing more than teaching leaders at the pinnacle of military organizations to deal with the perils of uncertainty by aligning organizations to environmental demands. For such institutions in a VUCA environment, this has a great deal of saleability.[60] Yet it bears a strong resemblance to some of the theories within the Business Community's Environmental Determined paradigm like Strategy-Structure Performance.[61] If one believed that all trends within the Business Community represented progress, this would mean that with the exception of the acceptance of 'New Science' thought, the Five American Armed Services lag behind. Yet the Five Armed Services and surrounding academic communities expend a significant amount of intellectual effort on considering the issues that face them, including the relationship between contemporary warfare and its effects on the levels of war. General Charles Krulak, then the Commandant of the Marine Corps, gave a speech at the National Press Club in October 1997. In this speech, he discussed two concepts. The first was the 'Strategic Corporal', a means to illustrate the requirement for low-level tactical leaders to be aware of the potential strategic ramifications of their decisions and actions. The second was the 'Three Block War', a means of explaining the requirement for leaders at all levels to be capable of ascending and descending the ladder of escalation quickly and changing the nature of operations from war fighting to peacekeeping/stability to humanitarian assistance and back again. General Krulak published two articles on the issue of the 'Strategic Corporal' and the concepts contained within became very popular in many circles.[62] The notion of the 'Strategic Corporal', however, was not without its critics. Guillot argued that the concept confused strategic ramifications with deliberate decision-making at the strategic level. The latter requires deeper analysis and evaluation, especially in terms of second- and third-order effects, whereas a 'Strategic Corporal' would not have the time to consider such things in anything approximating detail.[63] Another small article raised the point that if tactical decisions are so crucial under certain conditions, then the rank (and by inference, the experience) associated with particular layers of command should increase significantly to ensure that negative strategic ramifications are kept to a minimum.[64] Such debates have a profound effect on the understanding of the strategic level of war and a knock-on effect to strategic leadership.

**Development of Strategic Leaders**

Each of the Five American Armed Services deals with the issue of competencies associated with strategic leaders somewhat differently. Some patterns, however, will emerge from the examination of required competencies. Each service view is focused on the current situation, i.e. the VUCA environment and specific contemporary conflicts. This focus on the environment, of course, is the product of 'New Science' thought.

The U.S. Army has been the most prolific source of writing on the issue of required competencies. Leadership requirements, at all levels, are being addressed in the light of dealing with asymmetric warfare. One list of required competencies includes:

- Situational Awareness
- Strength of mind
- *Coup d'Oeil*
- Intelligent Risk-Taking
- Mental Readiness
- Knowing Yourself and Your Enemy
- Intellect
- Intuition
- Boldness
- Self-Reliance.[65]

Note that most of these are individual qualities and they convey an image of a heroic commander dealing with the anarchy imposed by an elusive opponent as opposed to a strategic leader dealing with an organization in war or peace. These traits could apply equally at the tactical and strategic levels. The U.S. Army War

# 5 CHAPTER

College's list of competencies infers a list of different, and more institutional, requirements for strategic leaders:

- Frame of Reference Development – identification of cause and effect in strategic environment
- Problem Management – systems thinking, pattern recognition, acceptance of ambiguity
- Political and social competence
- Consensus Building
- Negotiation
- Communication.[66]

Upon reading these terms, one starts to form a mental image of the institutional battlefields of Washington, DC, as opposed to an operational theatre. Yet the Army War College's list, despite its tone, could apply equally to either situation. It would not apply below the strategic level. While most of the competencies apply at lower levels, the requirement for consensus building does not. Strategic leaders, as those responsible for the organization, must be capable of building consensus and working with other organizations outside the formal chain of command.[67] This, on a bureaucratic battlefield, is tantamount to denying battle. It is, however, not well received within the ranks of organizations whose cultures thrive on the existence of a defined and recognized hierarchy such as the chain of command. Simply put, consensus building can be perceived as showing weakness, an inefficient way of carrying out one's duties, or both. A third list is aimed at developing leaders that can fight and win in a VUCA environment:

- Cognitive complexity
- Tolerance of ambiguity
- Intellectual flexibility
- Self-awareness
- Systems understanding
- Traditional leader qualities.[68]

# CHAPTER 5

This is a call to produce traditional leaders that have the emotional and mental capacity to cope with insufficient information over time. The Army War College has also sought to review strategic leadership competency requirements for the post-9/11 world based on the notion that the environment had become VUCA than hitherto. They identified the following requirements:

- Identity – the War College panel described this as: "...the ability to gather self-feedback, to form accurate self-perceptions, and to change one's self-concept as appropriate."[69]
- Mental agility
- Cross-cultural savvy
- Interpersonal maturity
- World Class warrior
- Professional astuteness.[70]

This list balances influence skills with cognitive and professional competencies. It could be interpreted as an attempt to reconcile the requirements of strategic leadership of an institution with the requirements of strategic leadership in war.

The U.S. Air Force's requirements are expressed far more simply. General 'Doc' Foglesong summarized the concept very well by stating that strategic leadership was really about having a 'big plan' and leading subordinates to its successful achievement.[71] The literature emanating from the Air Force community places the emphasis on ensuring that the 'big plan' rests on a foundation of truly strategic thought. This means that strategic thinking is the key element of strategic leadership, and it includes the use of deductive reasoning, the application of multiple frames of reference to any problem, convergent thinking, and conflict management.[72] The term 'convergent thinking' is a little confusing, as it is a combination of conceptualization and integration of multiple frames of reference.[73]

Both the U.S. Navy and Marine Corps appear to focus on leadership requirements in general as opposed to a single level. The U.S. Navy has a generic competency based model.[74] The USMC publication on leadership does not describe what competencies Marine strategic leaders require. This describes what Marine

# 5 CHAPTER

leaders require, regardless of the level of their employment. There is a set of leadership principles that prescribe a series of behaviours that Marine leaders are to develop.[75]

| Twenty-Eight Leadership Competencies | | | |
|---|---|---|---|
| **Leading Self** | **Leading Others** | **Leading Performance & Change** | **Leading The Coast Guard** |
| Accountability & Responsibility | Effective Communications | Conflict Management | Financial Management |
| Aligning Values | Team-Building | Customer Focus | Technology Management |
| Followership | Influencing Others | Decision-Making & Problem-Solving | Human Resource Management |
| Health & Well-Being | Mentoring | Management & Process Improvement | External Awareness |
| Self-Awareness & Learning | Respect for Others & Diversity Management | Vision Development & Implementation | Political Savvy |
| Personal Conduct | Taking care of people | Creativity & Innovation | Partnering |
| Technical Proficiency | | | Entrepreneurship |
| | | | Stewardship |
| | | | Strategic Thinking |

Table 4: Coast Guard Leadership Competencies[76]

The Coast Guard also has a broad list of competencies, and it must be said that this list is coherent with its definition of strategic leadership as 'Leading the Coast Guard'. Table 4 above provides a matrix that contains all 28 leadership

## CHAPTER 5

competencies. Note that it has grouped competencies around various layers of leadership, from the individual to others and from the abstract notion of change to the Coast Guard as an institution. It does not distinguish between knowledge and skills and contains a heady dose of managerial practice. This is due to the institutional focus of strategic leadership within the Coast Guard. Their vision of strategic leadership is clearer as they do not need to generate strategic leaders beyond the Coast Guard.

When it comes to strategic leader competencies, even the joint level has something to offer to the debate. The Industrial College of the Armed Forces, a subordinate institution of National Defense University, provides some instruction at the senior levels of the U.S. military on the topic of strategic leadership. The main document that supports this instruction is the *Strategic Leadership and Decision-Making Handbook (SLDM)*, which offers that strategic leaders must be capable of:

- Envisioning military roles to support policy objectives
- Envisioning military capabilities and programs to support policy objectives
- Developing consensus within an organization and with other organizations
- Ensuring commitment across the U.S. Government
- Program initiation

The sum of these requirements means that strategic leaders must have technical knowledge, interpersonal skills and conceptual skills.[77]

This list is based on an admixture of the NSS and NMS relationship as the basis of strategy and the institutional view. It could be argued that the *SLDM* is oriented to future American military leaders opposed to the future leaders of the five services.

### Is Strategic Leadership a Group or Individual Function?

Despite having organizations that place significant power in the hands of commanders, the Five American Armed Services are less prone to treating strategic

# 5 CHAPTER

leadership as an individual's responsibility. The literature suggests in most cases that strategic leadership is a team effort. The intellectual justification for this line of argument is the Environmentally Determined paradigm. Strategic leaders act as balancers of external and internal forces on organizations.[78] While they may be only one person, this task of guiding an organization internally and through the external environment (universally described in VUCA terms) forces it to be a collective effort.[79] Chilcoat's work on 'strategic art' makes a similar point by stating that true masters of strategic art are very rare, and as a result, strategic artistry is often shared within organizations and is dependent on interaction. The sophistication of this interaction governs organizational effectiveness.[80] Put another way, strategic leaders need to have mutual support with strategic thinkers and practitioners.[81]

Yet the Five American Armed Services produce individuals over time capable of forming strategic teams as opposed to building strategic teams over time. This is due to the tension extant within the Five American Armed Services between joint and service requirements, as: "The responsibility for developing leaders and leadership skills continue to reside in the services."[82] This simple observation may be somewhat innocuous by itself, but note that there is a significant difference between the leadership of a service or the entire military establishment. The services are preparing their leaders for the leadership of a service and its requirements with some attention to joint requirements. To make matters worse, strategic leaders are selected for having had a successful career at the tactical and operational levels, which means short-term results and a much-reduced requirement for the exercise of transformational leadership.[83] This means that strategic leaders may be selected for the wrong reasons. The most promising future strategic leaders might be selected from the best performers at general leadership, which are themselves selected from the best performers at direct leadership.

Each of the services has a similar approach to leader development and some are more explicit with regard to strategic leader development. This is due to the Chairman of the Joint Chiefs of Staff's (CJCS) recent direction to all services to have a common set of objectives in officer professional military education.[84] The CJCS' direction was for each educational institution dealing with intermediate and senior level education to incorporate both joint and single-service requirements

# CHAPTER 5

into their curricula. The Army uses a career-long pillars approach based on institutional education and training, experience through operational assignments and self-development. The Army War College curriculum is intended to prepare its students for future employment at the strategic level, and the emphasis has been placed on the strategic level as opposed to strategic leadership. The Air Force uses a 'Continuum of Education Framework' that teaches through a variety of methods: the profession of arms, military studies, international security, communications, and leadership. It also seeks to balance area expertise, assignments, training, deployments and mentoring. At the senior level, the Air War College is the primary institution for educating Air Force officers. Like the other services, the U.S. Navy relies on a career-long means of developing its leaders. It does this through operational assignments and institutional education as a preparation for the next assignment. This, in professional development terms, is a scheme of 'just-in-time' delivery. For strategic leaders in the U.S. Navy, the source of strategic leadership instruction is the National Security Decision-Making (NSDM) course at the Naval War College. Its curriculum is oriented towards decision-making at the senior levels through formulation and implementation of strategy as opposed to all of the dimensions of leadership. The contents of NSDM curriculum include the U.S. Army's FM 22-100, Henry Mintzberg's articles and books, and some of the other strategic planning literature.[85] The USMC's model is similar to the Army's, with the exception that career development is explicitly treated as a responsibility shared between the individual, their chain of command and the educational establishment in question. The strategic portion of Marine leader development has both military courses and civilian programs from the Center for Creative Leadership. The Coast Guard has similar practices to the Army and Marine Corps.[86] The balance between service and joint requirements for strategic leadership shows the tension that exists between the two definitions of strategy used within the Five American Armed Services.

## Part 4: Summary

The American military, on the surface, appears to have a more defined concept of strategy than the Business Community does. Closer examination revealed that two definitions were in use due to the nature of American civil-military relations.

# 5 CHAPTER

At the joint level, the concept of strategy is seen as the strategic level of war, which includes the requirement to bridge policy and military operations. This makes it a politico-military affair involving the ends-ways-means construct as opposed to a purely military affair. In order to avoid becoming overly political, the Armed Services focus on the institutional dimension of strategy, i.e. how to maintain and improve the situation facing the service. The American military's view of leadership varies from armed service to armed service, but this can be attributed to different service requirements with regard to 'direct' or 'general' leadership. Each service also has difficulty with the overlap between the concepts of command, leadership and management. The Five Armed Services have shown a preference for 'New Science' approaches to strategic leadership with the VUCA concept and acceptance of strategic leadership team as opposed to an individual. However, the tension created by three definitions of strategy (institutional, ends-ways-means, and the level of war) appears again with the requirement to centrally coordinate the curricula of the service institutions to ensure that a common standard of strategic level education is achieved.

# Chapter 6
## Comparing *Scylla* and *Charybdis*

This section will compare the differences between Charybdis (the Business Community) and Scylla (the Five American Armed Services) with regard to the definitions and concepts of strategy, leadership and strategic leadership. It will attempt to summarize the similarities and contrasts between the Business Community and the American military.

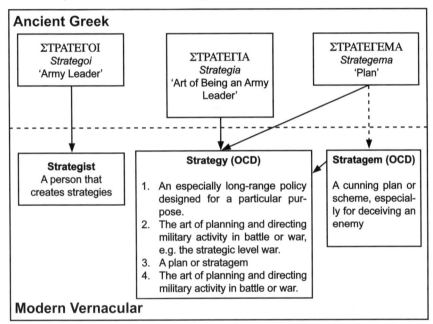

Figure 17: Conceptual change from the Ancient Greek to the Modern Vernacular[1]

## Part 1: *Scylla*'s Five Armed Services and *Charybdis*' Whirlpool of Business Thought on Strategy

The Business Community suffers from definitional anarchy whereas the Five American Armed Services only have to contend with three definitions of strategy

# 6 CHAPTER

(strategy as a level of war; strategy as a relationship between ends, ways, and means; and strategy as a long-range plan for the preservation and enhancement of an institution).  Both, however, face the challenge posed by the multiple connotations of the term 'strategy' and the general lack of use of the term 'stratagem' due to the existence of the third definition of strategy.  The latter provides an excellent means of separating the concept of strategy from its application in language, but it is seldom used in the military or commercial vernaculars.  As a result, the term 'strategy' is defined by its context.  It is remarkable that the Ancient Greeks had precise terms for the leader, the leader's art and the leader's plans.  Over time, the terms for the art and the plans have become the same word and all three terms have expanded beyond the military realm.  The term strategist has lost its purely military definition as depicted in Figure 17 (page 83).  The use of clear descriptors, i.e. corporate strategy or institutional strategy, would provide a lot of clarity as well as the renewed use of the term 'stratagem'.

## Part 2: *Scylla*'s Five Armed Services and *Charybdis*' Whirlpool of Business Thought on Leadership

The difference between the two versions of the same concept is that the Business Community has a simpler, albeit loose, definition of leadership being an influence process and management being an organizational process used by those in authority. In effect, the loose consensus within the Business Community is related to the unconscious acceptance that leadership is, by its very nature, *de facto*, whereas management is seen as *de jure* leadership.  This stems from the rise of 'New Leadership'.  It, however, connotes traditional management activities negatively. The Five American Armed Services have a different concept, command, which by its nature includes the conveyance of authority to a leader, the essence of *de jure* leadership, and a series of organizational activities.  Commanders exercise command over forces, which includes the exercise of both *de jure* and *de facto* leadership over subordinates, and staffs, which provide management functions and services, support them.

# CHAPTER 6

|  | *Charybdis* – the Business Community | *Scylla* – the Five American Armed Services |
|---|---|---|
| Strategy | Definitional Anarchy | 1) Level of War<br>2) Long-range plan for the preservation or enhancement of an institution |
| Leadership | Influence process within a framework of leadership and management | Influence process within a framework of command, leadership and management |
| Strategic Leadership | Subject to a debate between the proponents of the Environmentally Determined Paradigm and the Choice Driven Paradigm | 1) Institutional view<br>2) Acceptance of 'New Science' with VUCA |

Table 5: Comparison of the Five American Armed Services and the Business Community

## Part 3: *Scylla*'s Five Armed Services and *Charybdis*' Whirlpool of Business Thought on Strategic Leadership

Table 5 above provides a summary of the views within both camps on the concepts of strategy, leadership and strategic leadership. It is somewhat ironic that both camps have been focusing on the nature of the external environment. That, however, appears to be the *zeitgeist* of the early 21st century, and it has provided a fertile environment for the advocates of 'New Science' approaches to leadership. There is a difference between the American military and the Business Community apart from the debates within the Business Community between the Choice Driven and Environmentally Determined paradigms. The subtext to the

# 6 CHAPTER

military literature appears to recommend that leaders learn to become accustomed to VUCA whereas the subtext to the Business Community's literature appears to recommend that leaders seek to understand the world around them. The Business Community's premise is that the external realm can be understood whereas the latter suggests that while an understanding of the environment is desirable, it is certain to change, and therefore, such efforts may be wasted.[2]

# Chapter 7
## *Odysseus*: The Canadian Forces

The CF, in this monograph, is like Odysseus. Homer wrote, using Odysseus' voice:

> ...we sailed up the straits, wailing in terror, for on the one side we had Scylla, and on the other the awesome Charybdis sucked down the salt water in her dreadful way. When she vomited it up, she was stirred to her depths and seethed over like a cauldron on a blazing fire; and the spray she flung up rained down on the tops of the crags at either side. But when she swallowed the sea water down, the whole interior of her vortex was exposed, the rocks re-echoed to her fearful roar, and the dark blue sands of the sea-bed were exposed.
>
> My men turned pale with terror; and now, while all eyes were on Charybdis as the quarter from which we looked for disaster, Scylla snatched out of my ship the six strongest and ablest men. Glancing towards my ship, looking for my comrades, I saw their arms and legs dangling high in the air above my head. "Odysseus!" they called out to me in their anguish. But it was the last time they used my name. For like an angler on a jutting point, who casts his bait to lure the little fishes below, dangles his long rod with its line protected by an ox-horn pipe, gets a bite, and whips his struggling catch to land, Scylla had whisked my comrades, struggling, up to the rocks. There she devoured them at her own door, shrieking and stretching out their hands to me in their last desperate throes. In all I have gone through as I explored the pathways of the seas, I have never had to witness a more pitiable sight than that.[1]

# CHAPTER 7

## Part 1: Strategy in the Canadian Forces

### Strategy in the Politico-Military Context

The CF and American military have similar concepts of strategy but for very different reasons. The Five American Armed Services shy away from the definition of strategy as a level of war in order to stay in tightly defined politico-military parameters, e.g. the relationship between the NSS and the NMS. The Five Armed Services tend to use the definition of strategy as a long-range plan for the preservation and enhancement of an institution instead. The CF is different in that it has not normally had the benefit of an overt process and formal statement of grand strategy like the NSS.[2] As a result, it is necessary to venture into the realm of policy to set the context for a discussion of strategy.

Conventional wisdom holds that the Government of Canada, in terms of national security, has not been very good at providing clear policy direction. While this perceived paucity of direction has been a source of criticism, the situation has changed with the publication of the International Policy Statement (IPS).[3] Some may be tempted to offer the rebuttal that a Government White Paper on Defence is analogous to the NSS, but given that a White Paper on Defence only discusses defence policy as opposed to the broader grand strategy, they really represent a strategic document outlining the policies and tasks for the CF. As strategic documents, however, previous White Papers have had flaws that diminished their utility. The 1994 White Paper was oriented towards the maintenance of general-purpose combat capability, albeit in a resource-constrained era, but it failed to prioritize between competing efforts. Its predecessor, the 1987 White Paper, was lavish in terms of growth and acquisitions, but could not be sustained.

Donald Nuechterlein, an American political scientist, provided a useful model for analyzing national interests and for showing their relationship between the two streams of thought on international relations. Such interests provide a justification for actions, or a reason for why any given action was taken. Nuechterlein's model contains the four basic types of national interest:

# CHAPTER 7

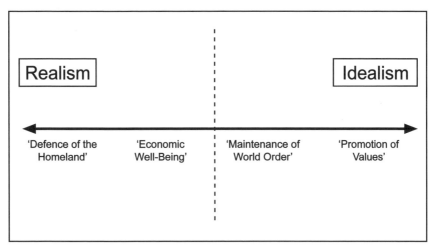

**Figure 18: National Interest Framework**[4]

Figure 18 above depicts the degree to which interests (Defence of the Homeland, Economic Well Being, Maintenance of World Order or the Promotion of Values) relate to a spectrum with Realism at one end or Idealism at the other.

It has been argued that Middle Powers like Canada have tended towards Idealist-based national interests. Their behaviour stems from altruism or notions of 'good international citizenship'. This, of course, leads to a preference for multilateral approaches to international conflict.[5] Also, they have shown a willingness to sacrifice national sovereignty to international organizations.[6] This is often the one and only means for Middle Powers to influence others to serve their own interests. This is where the implied message can cloud the reality of the situation. A policy of cooperation does not mean that national self-interest is non-existent in the Middle Powers. This fallacy rests on the assumption that only the powerful states can have interests. Contrary to popular belief, Canadian interests do not sit at the idealist end of the national interest spectrum. However, despite popular national belief, there is a fundamentally rational approach to the definition of Canadian national interests. There appears to have been a cost-benefit analysis applied in some instances that matched the collective belief. The two should not be confused with one and other.

# 7 CHAPTER

Canada has been privileged enough to have one of the most secure geopolitical positions in the world. It shares the North American landmass with the United States and since the demise of the Soviet Union, there has no tangible conventional military threat to Canada. The absence of conventional military threat has meant that there is little interest in security issues in Canada, as noted by Ross Graham, the Director General of Defence Research & Defence Canada Atlantic: "The domestic focus of most parliamentarians is consistent with the priorities of the Canadian public. A secure geopolitical situation has allowed politicians and the public the luxury of ignoring security matters."[7] Yet it would be misleading to state that this was not the case during the Cold War. Canada's military has been a prisoner of Canadian geography, and it has been: "...confronted by the implied requirement to defend the indefensible. Paradoxically, the vast size of the country works to favour defence, even as it renders it a virtually impossible job in traditional, physical terms."[8] The physical defence of Canada would be difficult to achieve without bankrupting the nation, so Canadians have become accustomed to accepting a certain level of risk, but this should also be tempered with the observation that the probability of a conventional military attack is extremely low. In short, the 'Defence of the Homeland' is not a significant concern.

The interest of economic well-being has been far more significant for Canada in the 1990s and has been a dominant theme for both the Mulroney and Chrétien governments. During that time, the Canadian government shaped its foreign policies toward an economic focus.[9] There were some other moments that suggested otherwise, such as the 1997 Ottawa treaty on landmines, but this was the well-publicized exception to the Canadian government's efforts to increase its volume of international trade. To wit, the numerous federally sponsored 'Team Canada' trade junkets in the 1990s suggested a far greater depth of interest than attempts at disarmament. There is an element of mutual support between economic well-being and the maintenance of world order.

Yet this interest cannot be separated from the Canadian relationship with the United States. The southern neighbour is Canada's greatest trading partner, and this is what the Canadian government has become sensitive to in the course of Canadian-American relations. Any efforts made to defend North America are

# CHAPTER 7

made in this light. For the United States, it is about defending the 'homeland', but for Canada, it is about preserving or enhancing the volume of trade with the United States.[10] Canadian prosperity is directly linked to harmonious relations with the United States.

This relationship with the United States has loomed large on the Canadian psyche, and it affected Canadian national interests significantly. Poll data taken in the mid-1990s indicated a: "...powerful streak of democratic moralism that pervades almost all of Canadians' thinking about international affairs..."[11] This penchant for democratic moralism creates a preference for multilateral approaches to conflict. Yet there is another reason for this preference. Canadian multilateral policies provide a means of balancing against the effects of the incredible power of Canada's closest ally.[12] This multilateral notion has been tied to 'cooperative security', and when associated with identity politics and a rights-based political culture, it is also a means of: "...reducing the burden that foreign policy imposes on domestic society..."[13] In short, a multilateral approach means that a Canadian voice, independent of the United States, may be heard. In addition, the cost of Canadian foreign policy will not be as heavy as it could be on Canadians.

In terms of national interests, Canada has shown a significant amount of will to demonstrate its virtues to others. It has been on a number of occasions that the case that foreign policy actions have been seen as a means of demonstrating Canadian values to others.[14] During the 1990s, Canadians began to view peace support operations in this light. What began with Lester B. Pearson's interference in the Suez crisis became:

> ...an *idealpolitik* distinguisher without undermining the European and North American defence roles... These missions also served as a handy justification for reductions in budget, force size and capabilities, as well as a convenient way to ensure that - while minimizing expenditures - Canada's forces operated as an extension of policy by other means to help 'retain a seat at the table'.[15]

# CHAPTER 7

Peacekeeping, while seen an excellent means of promoting Canadian values such as peace and order abroad was also a means of ensuring that Canada, as a middle power, had a greater voice in the international community.[16] It also allowed Canada to make direct military contributions to the broader western world by helping ensure that small 'brushfire' wars did not lead to a clash of superpowers.[17] The notion of the greater voice, however, should not be overlooked in terms of *realpolitik*. What was an expedient means of maintaining world order came to be seen as a means of promoting Canadian values abroad and the argument seemed to sway Canadian political elites. There is, however, a flaw with this line of argument. The means and the ends sometimes become confused. Peacekeeping suffers from this confusion. It has been argued that peacekeeping:

> ...appeals to Canadian popular self-images and sentiments of altruism and generosity... [and] represent[s] Canadian multiculturalism, tolerance and respect for the rule of law... Peacekeeping also meshes well with Canadian foreign policy conceptions of Canada as a ' middle power'... The desire to be both represented and consulted on international affairs is an important driving force behind Canadian foreign policy, and peacekeeping has helped maintain Canada's profile and influence as an independent sovereign actor in the world...[18]

The collective belief is not the source of the interest, and one ought to separate declaratory policy (e.g. promotion of Canadian values) from the actual interests (e.g. economic well being and the maintenance of world order).

A pattern of behaviour has emerged over time that allows for the identification of Canadian national interests. Canada has vital interests and has expressed how various national assets and resources, including the CF, will be assigned to achieve policy objectives related to those interests. Furthermore, the association of military means to political ends has been dominated by the depth of the interest in the maintenance of world order. The military instrument of power has been used incrementally to achieve the desired effect of being seen by Canadian allies to participate in sufficient measure.[19]

# CHAPTER 7

Canada's declaratory preference for idealist interests (namely the maintenance of world order and promotion of values) is at odds with the majority of the realist-oriented literature on strategy. This leads to the notion that Canada does not engage in grand strategic thought; such notions are false as its policies are based on a combination of realism and idealism as opposed to pure *realpolitik*.

The effect of a declaratory idealist-driven policy can be felt at the strategic and operational levels of war. It has created the conditions for the CF to focus on tactical doctrine and activity. This, however, has meant that Canadian military contributions to other operations have borrowed strategic or operational concepts from others, such as the United Kingdom and the United States. Cases where Canada operated as a single military entity, at those levels, are rather rare.[20] This has led some to conclude that for Canada, there is no true operational level as the strategy of making contributions to coalitions sees the achievement of strategic objectives through the mere presence of tactical forces.[21] The desired effect of such a stratagem is to preserve or improve Canada's position in the world order as opposed to a direct military effect. This, by no means, is meant to trivialize the hard work and sacrifices of Canadian soldiers, sailors and airmen, but rather to describe the context of the Canadian definition of the strategic level of war.

**Definition(s) of Strategy**

The CF definition of strategy is similar to the American definition in that it does not distinguish between policy and strategy. Strategy is defined as: "The application of national resources to achieve policy objectives".[22] The definition of the term used here is not purely military and goes beyond the 'bridging' of policy and military operations. The ends are provided and it is about matching ways and means. This raises the question of who actually determines Canadian strategy, as it is both a political and military issue. As the definition spans both, one would assume that the Prime Minister and Cabinet do this with support from the Privy Council Office. The National Defence Act does not explicitly mention strategy; it states that the Minister of National Defence is responsible for: " ...the management and direction of the Canadian Forces and of all matters relating to national defence..." while the Chief of the Defence Staff (CDS) is charged with:

# CHAPTER 7

"...the control and administration of the Canadian Forces."[23] This suggests that, at law, strategy is a government as opposed to military function, and this is consistent with the definition used by the CF. This should not be construed as stating that at law, the military has no role in strategy formulation and implementation. The integrated structure within National Defence Headquarters, although problematic in other ways, has been one means of ensuring that military advice informs the formulation and implementation of policy and strategy (as defined by the CF). The state of Canadian civil-military relations also affects the formulation and implementation of military strategy. Civilian control of the military in a *de jure* sense rests with the Prime Minister, the Minister of National Defence and Cabinet in that order. The exercise of *de facto* civilian control is done using at least one of two manners. The first is to exercise direct control over military policy and the second is to exercise control through resources. The former, especially in the Canadian context, can lead to civilians making what are military decisions and vice-versa, but this is being addressed through the transformation of the command and control of the CF. The latter is analogous to 'agency theory' or transactional leadership where subordinates are rewarded for 'working' (complying with the superior's requests) and punished for 'shirking' (failing to comply).[24]

The other means of balancing military advice and civilian control is the nature of the CDS' responsibilities as described as the 'control and administration' of the CF. This should be taken to mean that the CDS is responsible for the allocation of military resources and tasks in support of policy objectives. As a result, the draft publication *Leading the Institution* describes the CDS as a strategic commander.[25] Due to recent command and control changes, the CDS is a strategic leader of the institution and of the 'level of war'. Prior to the change, the CDS was primarily a strategic leader only in the institutional sense. The same publication attributes a different definition of strategy that is more in tune with the strategic level of war: "...the art of distributing and applying military means, or the threat of such action, to fulfil the ends of policy".[26] This definition is more precise in terms of the scale but also allows for the co-existence of two connotations of strategy. It ought to be used across the CF as the definition of strategy rather than the prior definition. These two also exist in the Five American Armed Services: strategy as a level of war and strategy as a long-range plan for the preservation or enhancement of an

# CHAPTER 7

institution. However, the institutional definition has taken primacy like it has in both the American military and the Business Community.[27]

**Dimensions of Strategy**

Discussions in the literature on Canadian strategy of its time horizon, scale and consequences are based on institutional concerns. Much of this comes in the form of programs and other management constructs. For example, one general officer sought to illustrate the strategic challenges of balancing present requirements with future developments in a resource-constrained environment.[28] To be clear, the present and future need to be defined in terms of the Force Planning Horizons:

- Force Planning Horizon 1 (1-5 Years) – this focuses on current capabilities
- Force Planning Horizon 2 (5-15 Years) – this focuses replacing current capabilities
- Force Planning Horizon 3 (10-30 Years) – acquiring new capabilities.[29]

The present excludes anything that cannot be influenced quickly. Note that this means from a strategic perspective, five years from now and today are no different. In terms of the factors, the doctrine, *Leading the Institution*, is itself focused on the institution and as a result, provides greater detail on internal factors. This includes the tension between the maintenance of a professional ideology within the CF and the: "...ideologies of managerialism and entrepreneurialism so influential in organizational governance."[30] Put in another manner, the CF is an organization based on the 'guardian' moral construct and the adoption of elements of the 'commercial' moral construct must be made very carefully to avoid corrupting the institution.[31] This is an implicit attempt to counter the negative effects of the integration of National Defence Headquarters in the 1970s and the drive for greater efficiency that followed in the wake of the defence cuts associated with the 1994 White Paper on Defence.

In summary, the CF faces similar issues with regard to the concept of strategy as does the American military and the Business Community. One must be careful

# 7 CHAPTER

to examine the Canadian use of strategy without falling victim to the collective belief of Canadian values at one end or unwarranted criticism about Canada not being particularly strategic in orientation the other end of the spectrum. The CF has two different definitions of strategy in use as shown in Figure 19 below, but for the purposes of maintaining healthy civil-military relations, focuses on strategy at the institutional level. The other definition of strategy is far too close to policy for CF comfort.

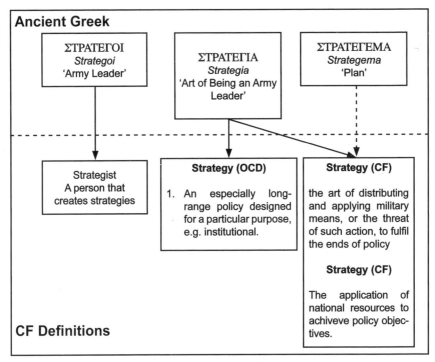

Figure 19: Definitions of Strategy in the CF

## Part 2: Leadership in the Canadian Forces

The CF doctrine on leadership uses a very inclusive definition of the concept. It is so inclusive that, were it not for the existence of specific criterion, it could be used as a definition of 'command'. Leadership, in the official CF doctrine, is defined as: " ...directing, motivating and enabling others to accomplish the mission professionally and ethically, while developing or improving capabilities

# CHAPTER 7

that contribute to mission success."[32] This definition requires leaders to simultaneously achieve their assigned tasks with developing and improving those around them. This helps foster a corporate culture of professional development and reminds leaders of their responsibilities for institutional stewardship. It also reminds the reader of the requirement to maintain professional ethos and apply sound ethics in the conduct of one's tasks, i.e. to not let the horror of war or the challenges of duty erode one's conduct to the point that they are inconsistent with institutional or Canadian values. This is similar to the USMC's ethos-driven approach to leadership doctrine. This again is an institutionally driven definition that indicates the existence of a significant level of concern for the maintenance of a collective Canadian military identity within Canadian society. Finally, the definition states that leaders 'direct, motivate and enable' others, which is not far off of 'command, lead and manage'.

Figure 20: Command, Leadership and Management in CF Doctrine[33]

The CF definition of leadership needs to be compared to the CF definitions of command and management in order to set the context. In addition to this, it must be acknowledged that the term 'management' has a negative connotation to many within the CF. First, command, like in the Five American Armed Services, rests on the notion of authority vested in an individual by a superior. The act of

**THE *SCYLLA* AND *CHARYBDIS* OF STRATEGIC LEADERSHIP**

# CHAPTER 7

commanding, in the official CF terminology, includes the definition of management shown in Figure 20 (page 97). Management is described as a rational, 'goal-driven' process of controlling resources, including one's subordinates.[34] This leads one to conclude that it describes the framework that establishes *de jure* leadership. The term 'management' developed a negative connotation within the CF as it was seen as unprofessional for a military institution. Integration led to the adoption of management practices and these came to be seen as the source of an erosion of military ethos. Further erosion occurred with the 1990s drive for greater efficiency associated with resource reductions. It must be recognized that the official CF definition of the term management and the unintended consequences of policy decisions that use management as justification, are in fact, separate from one and other. This is a negative effect stemming from the imprecise use of language. The CF definition of leadership, like in the Business Community, has been described as an influence process but includes both *de facto* and *de jure* forms of leadership.[35] It is very difficult for leaders at the senior level to exercise leadership in terms of personal attributes. They have been forced into adopting indirect or symbolic forms of leadership, which run the risk of being perceived as 'managerial behaviour' due to the level of indirect contact with their subordinates and the requirement to control large and complex entities.[36] Some, unfortunately, begin to compare the 'managerial behaviour' with their high expectations of the exercise of leadership and the senior leaders, regardless of their acumen, intelligence, or strong leadership abilities, are found wanting.

In an increasingly individualistic society somewhat hostile to authority and accustomed to a culture of entitlement, the maintenance of military values is crucial for the institution to remain capable of acting when so required. The CF military values are:

- Duty
- Loyalty
- Integrity
- Courage.[37]

# CHAPTER 7

The set of military values, like some of the American military services', is founded on the military's function to wage war when so ordered by government and military members have 'unlimited liability', i.e. they may have to lay down their lives in the course of duty. The values stress duty, the relationship with authority and the requirement for authority to both discipline and care for subordinates, which ought to see both leaders and subordinates bond into a team.

The definition of leadership is, unlike the Five American Armed Services, consistent across the CF. This institution faces some of the same challenges as its American counterparts in terms of overlapping concepts of command, leadership and management. A debate over the utility of particular theories of leadership in light of the maintenance of professionalism and suitable ethos is a recurring theme in the leadership doctrine. This suggests that while the CF remains a reflection of Canada, the institution requires that the values espoused by service personnel are not necessarily a direct reflection of Canadian values.

## Part 3: Strategic Leadership in the Canadian Forces

It is difficult to find an explicit definition of strategic leadership in the CF leadership-doctrine publications. There is a significant amount of implicit references to related concepts. The publication *Leading the Institution* provides the best example; it deals, as a whole, with the exercise of leadership over the CF. Consequently, the focus of strategic leadership uses the institutional definition as opposed to the level of war.[38] To be clear, this means strategic leadership has been considered a long-term (and largely managerial) endeavour, whereas the use of military forces in pursuit of national policy objectives has been bounded in time and space. Alternatively, this might mean the application of a particular stratagem in order to achieve national policy objectives.

**Concept & Definition**

The institutional nature of strategic leadership has been articulated clearly in publications like *Leadership: Conceptual Foundations* and the draft *Leading the Institution*. The aim is to ensure that the CF remains effective and can achieve

# 7 CHAPTER

its tasks through integration, adaptation and institutional health.[39] Institutional effectiveness is a combination of organizational and professional effectiveness.[40] Institutional effectiveness, as depicted in Figure 21 below, is a way of showing that the CF uses the set of means that are acceptable to Canadians and that still achieve the desired ends. The associated description of responsibilities (e.g. mission success, internal integration, care of members, external adaptability and ethos) is similar to the American military's.[41] This is not surprising given that the Five American Armed Services and the CF all view strategic leadership in the institutional sense. As mentioned earlier, the exercise of strategic leadership over an institution is seldom done face-to-face. Strategic leaders must exercise indirect and symbolic leadership.[42] Compliance with their demands is often borne of *de jure* vice *de facto* leadership as many subordinates do not in the course of their duties interact directly with the strategic leader. The latter, as a result, has to be the 'leader of leaders' and exercise both *de jure* and *de facto* leadership with mastery. Symbolic leadership includes public acts designed to communicate specific messages to subordinates. It is an indirect exercise of *de facto* leadership used to balance the effects of *de jure* leadership.

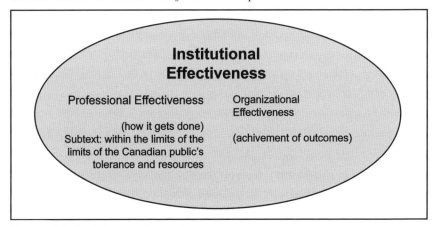

Figure 21: Institutional Effectiveness

The CF doctrine on strategic leadership attempts to make some compromises between the Five American Armed Services' views and the Business Community's views. It is similar to that American military literature in that it frames 'leading the institution' within a complex environment. The strategic leader needs to

# CHAPTER 7

understand the nature of this environment.⁴³ It is similar to the commercial literature in that it appears to prescribe that strategic leaders need an in-depth knowledge and understanding of the environment. The CF's environmental view is limited to what can be influenced; this means the 'institution environment' (as represented by the domestic, international and military environments).⁴⁴ For example, *Leading the Institution* devotes a chapter, titled 'Working the Town', to the environment in the federal government.⁴⁵ While the American military appears to prescribe 'get used to VUCA' and the Business Community appears to prescribe 'know your environment', the CF appears to prescribe 'know your *institutional* environment'.

**Is Strategic Leadership a group or individual function?**

The nature of the command and control arrangements of the CF has created some confusion as to the classification of various levels of command according to the levels of war. The second order effect of this is that it becomes less clear whether strategic leadership rests solely in the hands of the CDS or in those of a larger group that has been affected. The term 'strategic', for the Canadian Forces, has normally been associated with the CDS, the Vice CDS and the Environmental Chiefs of Staff (ECS) such as the Chief of Maritime Staff (CMS), Chief of Land Staff (CLS) and Chief of Air Staff (CAS). Such a definition was based on the requirement to translate White Paper direction (itself only a strategic document) into Defence Planning Guidance (another strategic document, although more closely aligned with the level of war definition of strategy).⁴⁶ This leads to subordinate layers of command self-identifying as operational if they were not a tactical unit. Given the recent changes created by Canadian Forces transformation, four force-employing headquarters actually function at the operational level, namely: Canada Command (Canada COM), Canadian Expeditionary Force Command (CEFCOM), Canadian Operational Support Command (CANOSCOM) and Canadian Special Operations Forces Command (CANSOFCOM). The ECS are responsible for force generation and do not employ forces. As a result, their responsibilities, such as the exercise over command and control of uncommitted forces, force generation for contributions to domestic or international operations and force development issues within a CF

# 7 CHAPTER

context, span multiple levels simultaneously and defy easy categorization into a single level of war. Strategic leadership, while vested in the CDS by the National Defence Act for the 'control and administration of the Canadian Forces', is a group activity. A *de facto* 'strategic body' consisting of the VCDS, the Chief of Military Personnel, the Chief of Defence Intelligence, the three ECS, and the Judge Advocate General directly supports the CDS.[47] Strategic leadership of the institution is a group activity as these senior military leaders provide advice to and participate in the CDS' planning efforts. Thus only a relatively small segment of the CF engages in strategic leadership.

**Development of Strategic Leaders**

Strategic leaders, according to the CF doctrine, require particular competencies. These are 'leader elements':

- Expertise (Strategic level of war)
- Cognitive Capacities (creative abstract)
- Social Capacities (Inter-institutional)
- Change Capacities (Paradigm Shifting)
- Professional Ideology (Stewardship).[48]

These elements are similar to the American military's sets of requirements in that they list the professional expertise, the cognitive skills, the need to lead change, the need to work well with those outside the institution and to maintain the ethos of the institution. However, they could make the requirement to communicate effectively more explicit. Strategic leaders must rely on: "...secondary and tertiary influence processes for the senior leader to communicate institutional priorities across organizational systems."[49] This is a key competency for leaders in general, but given that strategic leaders must employ indirect leadership as their main tool, the ability to articulate concepts and intent clearly across multiple subordinate layers of command is crucial.

There is a fundamental decision for any military institution associated with professional development; should professional development prepare the leader

# CHAPTER 7

for the next job or prepare all leaders over time for the penultimate job? On one hand, preparing the leader for the next job places an institutional emphasis on leaders gaining experience, but on the other preparing them over time, places the emphasis on individual education and training. The American military appears to favour preparing personnel for the next job, and the Business Community doesn't tend to favour any particular approach. One of the challenges for both approaches is that the CF structure has many bureaucratic jobs, i.e. they maintain the structure of the institution but do not necessarily require a military member to carry them out, in addition to the purely militarily professional jobs.[50] The bureaucratic jobs, while necessary to maintain an institution, create artificial training requirements (i.e. that geared to teaching CF personnel how to live within a bureaucracy) or provide too much of one kind of experience for a military professional (i.e. military skills, fitness, and ethos erode in an 'office' environment). The CF uses a series of professional development 'pillars' (Education, Experience, Individual Training, Self-Development) to describe the various methods of developing personnel. The challenge ahead is how to guarantee the optimal combination to develop strategic leaders in light of:

- Short term requirements – this includes domestic and international operations
- Prevailing culture and ethos – that of a tactically oriented institution concerned with the preservation of military virtues in an increasingly individualistic society
- PD Pillars / Training Regimens – time spent in individual training and education means time is not spent in acquiring experience.[51]

This is significant as the time allotted to educate senior leaders is eats into the time for them to gain experience.

At present, the CF professional development system prepares leaders for the next tier of positions. Individual training and education efforts are devoted to preparing the leader for a series of jobs associated with their rank(s), which represents a compromise between the two approaches to leader development as a series of jobs, of which only some are purely military, prepares the leader for

# CHAPTER 7

the next rank. At the general officer level, where individuals deal in the realm of strategy, there are two courses: the Advanced Military Studies Program (AMSP) and the National Security Studies Program (NSSP). The AMSP is intended to prepare Colonels and Captains (Navy) for service at the operational level, and the NSSP is to prepare general officers and selected others for: "...strategic leadership responsibilities in the development, direction and management of national security and defence policy."[52] It consists of several subordinate courses, including a course on strategic command and executive leadership. This course consists of three modules as follows:

a. Command and Leadership (COM). This module is concerned with the examination of the nature, legitimacy and structure of command at the strategic level. In addition, the module examines command and control systems and their vulnerabilities in contemporary and in predicted future circumstances; the distinction between command and control will be considered from a theoretical point of view. This module additionally considers transformational leadership concepts, theories and includes self-assessment feedback to enhance individual personal leadership styles and behaviour.

b. Ethics and the Military Profession (EMP). The three interlinked concepts of ethos, ethics and professionalism will be examined in a military context that will focus upon issues related to authority, accountability and responsibility. Course members will review the fundamentals of moral philosophy and ethical reasoning; they will explore some of the major ethical issues associated with their profession, particularly at the senior executive level. They will further consider the challenges of inculcating ethical behaviour in their subordinates, peers and superiors.

c. Communications and Media Relations (CMR). This module is concerned with the development of strategic communications plans, the application of crisis public relations management, and

# CHAPTER 7

the understanding of the military-media relationship. Examining public attitudes on domestic and international issues concerning the Canadian Forces, it will enable course members to prepare communications strategies and apply public affairs resources.[53]

This is the only course devoted to strategic leadership, and there is an equivalent module dedicated to management. Given the quantity of strategic leaders in the CF, the quantity and timing of the training is suitable for the CF's requirements. There is a challenge, however, from the emphasis on management within the NSSP curriculum as it is at odds with the leadership doctrine.

## Part 4: Reconciling the Canadian Forces with the Two Sources of Strategic Leadership Thought

So where is the CF exactly? Table 6 (page 106) summarizes its position on the three concepts of strategy, leadership and strategic leadership. In terms of strategy, the Business Community represents definitional anarchy, that is, there is no consensus upon which of the hundreds of variations of definitions in use, including the OCD definitions. Not surprisingly, the CF and Five American Armed Services think alike about the concepts of strategy. It is viewed either as a long-range plan to preserve or enhance the institution or as the level of war that bridges policy and military operations.

In terms of leadership, the CF is closer to the American services than it is to the Business Community, but that two allied militaries would have similar doctrines is not unexpected. Both have to wrestle with the overlapping concepts of command, leadership and management, but the CF shares something in common with the Business Community. Both the Officer and NCO corps alike see the term 'management' in a somewhat negative light, although there is a difference between CF doctrine and the views from within. Both the CF and the Business Community treasure the notion of leadership as an influence process; the key difference is that the concept of command allows for the CF to accept that the influence process occurs within an organizational hierarchy.

# 7 CHAPTER

In terms of strategic leadership, the CF tends to be more like its American counterparts. While a debate continues to occur within the ranks of the Business Community between the proponents of the Environmentally Determined and Choice Driven paradigms, it has not occurred to the same extent in the two militaries. The institutional view of strategic leadership, based on the institutional definition of strategy, is the mainstream view for both militaries. They differ, however, in that the Five American Armed Services appear to be more receptive to 'New Science' and related concepts whereas the CF sees the concepts of VUCA and the Environmentally Determined paradigm through an institutional prism. For the CF, strategic leadership is all about the institution and less about leading at the strategic level of war.

|  | *Charybdis –* the Business Community | *Odysseus –* the Canadian Forces | *Scylla –* the Five American Armed Services |
|---|---|---|---|
| Strategy | Definitional Anarchy | 1) Level of War 2) Long-range plan for the preservation or enhancement of an institution | 1) Level of War 2) Long-range plan for the preservation or enhancement of an institution |
| Leadership | Influence process within a framework of leadership and management | Process of directing, motivating and enabling others within a framework of command | Influence process within a framework of command, leadership and management |
| Strategic Leadership | Subject to a debate between the proponents of the two paradigms | Institutional view with a blending of 'VUCA' and the Environmentally Determined Paradigm | 1) Institutional view 2) Acceptance of 'New Science' with 'VUCA' |

Table 6: The Business Community, the CF and the Five American Armed Services Compared

# CHAPTER 7

What is interesting and may be an avenue for the CF to explore is why both the American and Canadian military services seem to prefer Environmentally Determined Paradigms of strategy to choice driven paradigms. It begs the question of why a debate still occurs within the Business Community but not in the Canadian and American military institutions.

The answer to the question is simple. Debates still occur within that community due to the imprecise use of the term 'strategy', the relative absence of the term 'stratagem' and the current trend of promoting *de facto* leadership at the expense of *de jure* leadership. While it is true that the American and Canadian military institutions do not face such debates, they also face similar challenges with regard to the terminology and the promotion of one form of leadership over the other. What is needed is clear thought and language on strategy and a more balanced view of the two forms of leadership on everyone's part.

# ENDNOTES

# Endnotes

### Chapter 1  *Scylla* and *Charybdis*

1   Homer, *The Odyssey*, (Toronto: Penguin Books, 2003), p. 159. See also: Neil Phillip, *Myths and Legends*, (London: Dorling Kindersley, 1999), pp. 336, 1023 and 1039. This myth is thought to be the origin of the phrase 'between a rock and a hard place'.
2   The author is indebted to Dr. Andrew Godefroy for the concept.
3   For a thought-provoking discussion on the adoption of new ideas and their unintended consequences, see Jane Jacobs, *Systems of Survival: A Dialogue on the Moral Foundation of Commerce and Politics*, (New York: Vintage Books, 1992).
4   For a discussion of 'New Science', see p. 38 above.

### Chapter 2  Strategy in the Western World from Ancient Greece to the late Industrial Revolution

1   R.G. Collingwood, *An Essay on Philosophical Method*, (Bristol: Thoemmes Press, 1995), p. 204, cited in Jon Tetsuro Sumida, "The Relationship of History and Theory in *On War*: the Clausewitzian Ideal and its implications", *Journal of Military History*, Vol. 65, No. 2 (April 2001), pp. 333-354.
2   For the OCD definitions, see p. 22 above.
3   Robert M. Grant, *Contemporary Strategy Analysis*, (Malden: Blackwell, 1998), pp. 14-15. It is hoped that this work will be a furtive first step toward establishing such a theory.
4   Examples from the 1990s include Michael V. Harper and Gordon R. Sullivan, *Hope Is Not A Method: What Business Leaders Can Learn From America's Army*, (New York: Times Books, 1996), and William G. Pagonis with Jeffrey L. Cruickshank, *Moving Mountains: Lessons In Leadership And Logistics From The Gulf War*, (Cambridge: Harvard University Press, 1992). This leads one to hypothesize that military methods are more marketable in the periods of time following military victories. This is a potential avenue for future research.
5   The development of the *Strategoi* was part of Cleisthenes' 508 BC reforms of Athens. See: Stephen Cummings, "Pericles of Athens – Drawing from the Essence of Strategic Leadership", *Business Horizons*, Vol. 38 No. 1 (January-February 1995), pp. 23-24, Carnes Lord, "Leadership and Strategy", *Naval War College Review*, Vol. LIV, No. 1 (Winter 2001), pp. 139-145, and Daniel Moran, "Strategy", in Richard Holmes, Ed., *Oxford Companion to Military History*, (Oxford: Oxford University Press, 2001), p. 879.
6   Roger S.O. Tomlin, "Vegetius", in Holmes, Ed., p. 946. Vegetius' *De Re Militari* (The Military Institutions of the Romans) was divided into five volumes. The first two dealt with the recruiting and training of soldiers and the organization, officers and formation of a legion. The last two volumes dealt with fortifications and naval operations. The third volume was a compendium of tactics and strategy. See: Flavius Vegetius Renatus, "The Military Institutions of the Romans (*De Re Militari*)", in Major Thomas R. Phillips, U.S. Army, Ed., *Roots of Strategy*, (Harrisburg: Military Service Publishing Company, 1940), pp. 65-175.

# ENDNOTES

7 Previous scholarship on this issue held that strategy did not exist during this period, as there was no links between rulers' policies and military actions. For example, see Hans Delbrück, *History of the Art of War, Volume III, Medieval Warfare*, (Lincoln: University of Nebraska Press, 1990), pp. 323-328 and pp. 635-642. More recently, others have argued that this was not the case. For example, see: Malcolm Barber, "Medieval Strategy", in Holmes, Ed., p. 888. See also: Kelly DeVries, "The Art of Warfare in Western Europe during the Middle Ages from the Eighth Century to 1340", Book Review, *Journal of Military History*, Vol. 62, No. 1 (January 1998), pp. 193-194, and Sean McGlynn, "The myths of medieval warfare", *History Today*, Vol. 44 (January 1994), pp. 28-34.

8 Lynn Montross, *War Through the Ages*, Third Edition, (New York: Harper, 1960), p. 25.

9 Phillip Bobbitt, *The Shield of Achilles: War, Peace and the Course of History*, (New York: Anchor Books, 2003), pp. 95-204.

10 The terms 'strategy of annihilation' (*Niederwerfungsstrategie*) and 'strategy of exhaustion' (*Ermattungsstrategie*) originated with the German military historian Hans Delbrück. See: Hans Delbrück, *History of the Art of War, Volume I, Warfare in Antiquity*, (Lincoln: University of Nebraska Press, 1990), pp. 135-136 and Hans Delbrück, *History of the Art of War, Volume II, The Barbarian Invasions*, (Lincoln: University of Nebraska Press, 1990), pp. 379-382. See also: Gordon A. Craig, "Hans Delbrück: The Military Historian", in Paret, Ed., pp. 341-344. For a succinct description, see: Antulio J. Echevarria II, *Toward An American Way Of War*, (Carlisle: Strategic Studies Institute, 2004), p. 20.

11 Captain (Navy) Gordon Peskett, "Levels of War: A New Canadian Model to Begin the 21st Century", in Allan English, Daniel Gosselin, Howard Coombs and Laurence M. Hickey, Eds., *The Operational Art: Canadian Perspectives – Context and Concepts*, (Kingston: Canadian Defence Academy, 2003), pp. 97-127.

12 Howard attributes this to the fact that nineteenth century writing on strategy was influenced heavily by Napoleon's exploits. He cautioned, however, that this does not in any way reduce the importance of sound logistical plans. See: Michael Howard, "The Forgotten Dimensions of Strategy", *Foreign Affairs*, Vol. 57, No. 5 (Summer 1979), pp. 975-976.

13 Karl Von Clausewitz, *On War*, (New York: Modern Library, 1943), Book II, Chapter I, p. 62.

14 Definition cited in B.H. Liddell Hart, *The Strategy of the Indirect Approach*, (London: Faber & Faber, 1946), pp. 156-157. See also: Lieutenant Colonel Bill Bentley, Canadian Forces, *Professional Ideology and the Profession of Arms in Canada*, (Toronto: Canadian Institute of Strategic Studies, 2005), p. 21.

15 Colin Gray, *Modern Strategy*, (Oxford: Oxford University Press, 1999), pp. 5, 17, and 43, and Sir Michael Howard, "Grand Strategy in the Twentieth Century", *Defence Studies*, Vol. 1, No. 1 (Spring 2001), p. 1. The requirement to make national sacrifices is best expressed through the economist's metaphor of choosing between 'guns and butter'.

16 Moran, in Holmes, Ed., p. 881.

17 For the purposes of this work, the terms 'policy' and 'grand strategy' are considered to be interchangeable. Michael Howard argued that: "In the West the concept of 'grand strategy' was introduced to cover those industrial, financial, demographic, and societal aspects of war that have become so salient in the twentieth century. . ." in "Forgotten Dimensions", p. 975.

18   The use of campaigns to achieve particular goals has, over time, come to be seen as the hallmark of 'Operational Art'. See: B-GL-300-001/FP-001 *OPERATIONAL LEVEL DOCTRINE FOR THE CANADIAN ARMY*, (Kingston: Army Publishing Office, 1998), pp. 48-107, English et al., and B.J.C. McKercher and Michael Hennessy, Eds., *The Operational Art: Developments in the Theories of War*, (Westport: Praeger, 1996). Critics of this approach would argue that the revival of the concept of Operational Art came at a time when the U.S. Army was seeking intellectual justification for teaching its general officer corps how to be generals in the post-Vietnam reforms.

19   See: Lawrence Freedman, *Strategic Defence in the Nuclear Age*, Adelphi Paper 226, (London: IISS, 1987), p. 34, Michael Howard, "The Forgotten Dimensions of Strategy", *Foreign Affairs*, Vol. 57, No. 5 (Summer 1979), p. 975, and Paul Kennedy, "Grand Strategy in War and Peace: Towards a Broader Definition", in Paul Kennedy, Ed., *Grand Strategies in War and Peace*, (Cambridge: Yale University Press, 1991), pp. 2-4.

20   Bentley, p. 68.

21   Richard K. Betts, "Is Strategy an Illusion?", *International Security*, Vol. 25, No. 2 (Fall 2000), pp. 7 and 49.

22   For further details, see p. 31 above.

23   Andre Beaufre, *An Introduction to Strategy*, (London: Faber and Faber, 1963), p. 22, cited in Gray, p. 18.

24   Leland H. Jenks, "Early Phases of the Management Movement", *Administrative Studies Quarterly*, Vol. 5, No. 3 (December 1960), pp. 421-447 and Yehouda Shenhav, "From Chaos to Systems: The Engineering Foundations of Organization Theory, 1879-1932", *Administrative Studies Quarterly*, Vol. 40, No. 4 (December 1995), pp. 557-585. See: Frederick W. Taylor, *The Principles of Scientific Management*, (New York: Norton, 1911).

25   Charles Perrow, "The Short and Glorious History of Organizational Theory", *Organizational Dynamics*, Vol. 2 No. 1 (Summer 1973), pp. 4-5.

26   'Scientific Management' found favour in the Soviet Union in the 1930s.

27   Brian Leavy, "On Studying Leadership in the Strategy Field", *Leadership Quarterly*, Vol. 7, No. 4 (Winter 1996), pp. 435-436. See also: Kenneth R. Andrews, *The Concept of Corporate Strategy*, Revised Edition, (Homewood: Richard D. Irwin, Inc., 1980), H. Igor Ansoff, *The New Corporate Strategy*, (New York: John Wiley & Sons, 1965), Alfred A. Chandler, Jr., *Strategy and Structure: Chapters in the History of American Industrial Enterprise*, (Cambridge: MIT Press, 1962) and Alfred Sloan, *My Years with General Motors* (New York: Bantam Dell, 1963).

28   Katherine Barber, Ed., *The Oxford Canadian Dictionary*, (Don Mills: Oxford University Press, 1998), p.1435.

29   Definitions of the modern vernacular can be found in the OCD, p. 1435.

30   The source of such thinking is largely due to 'New Science'. For details, see p. 38 above.

## Chapter 3   Leadership in the Western World from Ancient Greece to the mid-20th Century

1   Lesley Prince, "Eating the Menu Rather than the Dinner: Tao and Leadership", *Leadership*, Vol. 1, No. 1 (2005), pp. 105-106 and 111.

# ENDNOTES

2   Richard A. Barker, "The nature of leadership", *Human Relations*, Vol. 54, No. 4 (2001), p. 476.
3   Max Depree, *Leadership Jazz*, (Toronto: Currency Doubleday, 1992), p. 140.
4   Joseph C. Rost, *Leadership for the Twenty-First Century*, (Westport: Praeger, 1993), pp. 37-38.
5   Felix Gilbert, "Machiavelli: The Renaissance of the Art of War", in Peter Paret, Ed., *Makers of Modern Strategy: from Machiavelli to the Nuclear Age*, (Princeton: Princeton University Press, 1986), pp. 12-15. By the late 18th and early 19th Centuries, this would return to being an increasingly political responsibility with concepts like the '*Levée en Masse*' or the 'nation in arms'.
6   Barker, p. 476.
7   Niccolò Machiavelli, *The Prince*, (Toronto: Penguin Books, 2003), pp. 79-80.
8   Brian Leavy, "On Studying Leadership in the Strategy Field", *Leadership Quarterly*, Vol. 7, Issue 4 (Winter 1996), pp. 435-436. See also: Colonel Bernd Horn, "Executive Leadership", in Colonel Bernd Horn, Ed., *Contemporary Issues in Officership*, (Toronto: CISS Press, 1996), pp. 123-124 and Rost, pp. 3, 6, 10, 14 and 16.
9   Gary Yukl, *Leadership in Organizations*, 4th Edition, (Upper Saddle River: Prentice-Hall, 1998), p. 409.
10  Rost, p. 27. The term 'Newtonian science' refers to the method of reducing any phenomena to its component parts to aid in explanation and / or comprehension. The term 'rational' refers to describe a decision calculus based on the deliberate maximization of benefits relative to cost. The term 'utilitarian' refers to a philosophy based on the notion of the greatest good for the greatest number.
11  Rost, p. 47. For a discussion of the early twentieth century thought on management, e.g. 'scientific management', see p. 21 above.
12  Rost, p. 24.
13  Robert J. House and Ram N. Aditya, "The Social Scientific Study of Leadership: *Quo Vadis*?", *Journal of Management*, Vol. 23, No. 3 (1997), pp. 409-421.
14  The use of the terms *de jure* (Latin: based on law or principle) and *de facto* (Latin: based on fact or practice) dimensions of leadership came in a conversation with Howard Coombs, 7 November 2006. See also: Howard Coombs, *Dimensions of military leadership: The Kinmel Park Mutiny of March 4/5, 1919*, CFLI Contract Research Report #CR02-0623, (Kingston, ON: Canadian Forces Leadership Institute, 2003), Ross Pigeau and Carol McCann, "What is a Commander?", in Bernd Horn and Stephen J. Harris, Eds., *Generalship and the Art of the Admiral: Perspectives on Canadian Senior Military Leadership*, (St. Catharines: Vanwell, 2001), pp. 83-97, and J.E. Adams-Roy, *The Role of the Lawful Order in Military Leadership: Necessary but Insufficient......or Insufficient but Necessary?*, CR01-0147, (Kingston, ON: Canadian Forces Leadership Institute, 2002). The official CF definition of leadership is: " *...directly or indirectly influencing others, by means of formal authority or personal attributes, to act in accordance with one's intent or a shared purpose.*" This definition as shown above appears in *Leadership in the Canadian Forces: Doctrine*, A-PA-005-000/AP-003, (Kingston: Canadian Defence Academy, 2005), p. 3.

# ENDNOTES

**Chapter 4** *Charybdis*: **The Swirling Vortex of Business and Social Science Literature**

1. John Storey, "What Next for Strategic-Level Leadership", *Leadership*, Vol. 1 No. 1 (2005), pp.89-90.
2. Giovanni Gavetti and Daniel A. Leventhal, "The Strategy Field from the Perspective of *Management Science*: Divergent Strands and Possible Integration", *Management Science*, Vol. 50, No. 10 (October 2004), pp. 1309-1311.
3. Bruce D. Henderson, "The Origin of Strategy", in Harvard Business Review, *Strategy: Seeking and Securing Competitive Advantage*, (Cambridge: Harvard Business Press, 1991), pp. 3-11 and Alfonso Montuori, "From Strategic Planning to Strategic Design: Reconceptualizing the Future of Strategy in Organizations", *World Futures*, Vol. 59, No. 1 (2003), p. 11.
4. Moshe Farjoun, "Towards an Organic Perspective on Strategy", *Strategic Management Journal*, Vol. 23 (2002), p. 561-563.
5. Henry Mintzberg, "The Strategy Concept II: Another Look at Why Organizations Need Strategies", *California Management Review*, Vol. 30 No. 1 (Fall 1987), pp. 25-26.
6. Mintzberg, "The Strategy Concept II", pp. 28-30.
7. Pol Herrmann, "Evolution of Strategic Management: The need for new dominant designs", *International Journal of Management Reviews*, Vol. 7, Issue 2 (June 2005), p. 112.
8. The definition of paradigm used in this monograph comes from Thomas Kuhn, *The Structure of Scientific Revolutions*, 3rd Edition, (Chicago: University of Chicago Press, 1996), p. 10. Kuhn stated that paradigms were novel enough to attract a lasting group of supporters away from competing ideas and that they were broad enough so as to leave a series of other problems for the supporters to address.
9. L.J. Bourgeois III, "Strategic Management and Determinism", *Academy of Management Review*, Vol. 9 No. 4 (1984), pp. 586-596.
10. See: Kenneth R. Andrews, *The Concept of Corporate Strategy*, Revised Edition, (Homewood: Richard D. Irwin, Inc., 1980), and Alfred A. Chandler, Jr., *Strategy and Structure: Chapters in the History of American Industrial Enterprise*, (Cambridge: MIT Press, 1962). For a discussion on both models, see: Robert A. Burgelman, "A Model of the Interaction of Strategic Behaviour, Corporate Context, and the Concept of Strategy", *Academy of Management Review*, Vol. 8, No. 1 (1983), pp. 61-70. Andrews was the inventor of Strengths-Weaknesses-Opportunities-Threats (SWOT) analysis to help organizations adapt themselves to organizational demands. See also: Rainer Feurer and Kazem Chaharbaghi, "Strategy development: past, present and future", *Management Decision*, Vol. 33, No. 6 (1995), p. 12.
11. Burgelman, pp. 62-63.
12. For an excellent summary of the various approaches, see: Henry Mintzberg, Bruce Ahlstrand and Joseph Lampel, *Strategy Safari: A Guided Tour through the Wilds of Strategic Management*, (Toronto: The Free Press, 1998), pp. 354-359.
13. See: H. Igor Ansoff, *The New Corporate Strategy*, (New York: John Wiley & Sons, 1965).
14. Montuori, p. 5.
15. Sandra L. Williams, "Strategic planning and organizational values: links to alignment", *Human Resource Development International*, Vol. 5, No. 2 (2002), p. 219.

# ENDNOTES

16    This was a means to unify all strategic nuclear target lists in 1960 and its use carried on into the 1980s. See: Desmond Ball and Jeffrey Richelson, Eds., *Strategic Nuclear Targeting*, (Ithaca: Cornell University Press, 1986).

17    Herrmann, p. 114. Interestingly, in the realm of political science, due to post-Vietnam disillusionment, there was a brief period in the early 1970s where 'bureaucratic politics' reigned supreme before the rationalist theories returned. See: Betts, p. 7. The term 'bureaucratic politics' originated with Richard Neustadt's book, *Presidential Power*, an examination of decision-making within the Kennedy Administration, but became well known due to a study of the Cuban Missile Crisis that offered three different explanations as to the logic of the decisions made during that crisis. The third explanation was based on 'bureaucratic politics'. See: Graham Allison, *Essence of Decision: Explaining the Cuban Missile Crisis*, (Boston: Little, Brown and Co., 1971), pp. 162-181.

18    See: Michael Porter, *Competitive Strategy: Techniques for Analyzing Industries and Competitors*, (New York: The Free Press, 1980) and Michael Porter, "How competitive forces shape strategy", *The McKinsey Quarterly*, Issue 2 (Spring 1980), pp. 34-50.

19    Boal and Hooijberg, pp. 521-522. For a study of its application to American civil-military relations, see: Peter D. Feaver, *Armed Servants: Agency, Oversight and Civil-Military Relations*, (Cambridge: Harvard University Press, 2003).

20    For a succinct discussion, see: Robert E. Hoskisson, Michael A. Hitt, William P. Wan and Daphne Yiu, "Theory and research in strategic management: Swings of a pendulum", *Journal of Management*, Vol. 25, No. 3 (1999), pp. 417-456. See also: Farjoun, pp. 561 and 564.

21    David J. Collis and Cynthia A. Montgomery, "Competing on Resources: Strategy in the 1990s", *Harvard Business Review*, Vol. 73, No. 4 (July-August 1995), p. 119.

22    This is not an isolated phenomenon; the five American armed services use it as well.

23    Richard A. Johnson, Fremont E. Kast, and James E. Rosenzweig, "Systems Theory and Management", *Management Science*, Volume 10, Number 2 (January 1964), p. 367. See also: Kenneth E. Boulding, "General Systems Theory – The Skeleton of Science", *Management Science*, Volume 2, Number 3 (April 1956), pp. 197-208 and Gregory A. Daneke, "The Reluctant Resurrection: New Complexity Methods and Old Systems Theories", *Journal of Public Administration*, Vol. 28, No. 1/2 (2005), pp. 89-106.

24    Gary M. Grobman, "Complexity Theory: A New Way to Look At Organizational Change", *Public Administration Quarterly*, Vol. 29, No. 3/4 (Fall/Winter 2005), pp. 356-358. Early works on systems theory include: Ludwig von Bertalanaffy, "General System Theory: A New Approach to Unity of Science", *Human Biology*, Vol. 23 (December 1951), pp. 303-361 and Kenneth Boulding, "General Systems Theory – the Skeleton of Science", *Management Science*, Vol. 2, No. 3 (April 1956), pp. 197-208. For a criticism of early applications of systems theory to organizational theory, see: Donde P. Ashmos and George P. Huber, "The Systems Paradigm in Organizational Theory: Correcting the Record and Suggesting the Future", *Academy of Management Review*, Vol. 12, No. 4 (1987), pp. 607-621. See also: Jeffrey Vancouver, "Living Systems Theory as a Paradigm for Organizational Behaviour: Understanding Humans, Organizations and Social Processes", *Behavioural Science*, Vol. 41, Issue 3 (July 1996), pp. 164-204.

# ENDNOTES

25  Ludwig von Bertalanffy, "The History and Status of General Systems Theory", *Academy of Management Journal*, Vol. 15, No. 4 (December 1972), p. 412. See also: Debora Hammond, "Exploring the Genealogy of Systems Thinking", *Systems Research and Behavioral Science*, Vol. 19, No. 5 (September-October 2002), pp. 429-439.

26  Patricia H. Werhane, "Moral Imagination and Systems Thinking", *Journal of Business Ethics*, Vol. 38, No.1/2, (June 2002), p. 33.

27  Gordon R. Sullivan and Michael V. Harper, *Hope is Not A Method*, (New York: Broadway Books, 1996), p. 29.

28  For example, see: Margaret Wheatley, *Leadership and the New Science: Learning about Organization from an Orderly Universe*, (San Francisco: Berrett-Koehler Publishers, 1992) and Margaret J. Wheatley and Myron Kellner-Rogers, "Self-Organization: The Irresistible Future of Organizing", *Strategy & Leadership*, Vol. 24, No. 4 (July-August 1996), pp. 18-25.

29  For further discussion of the concept of a vision, see p. 49 above.

30  See: Robert J. Mockler, "Prescription for disaster: failure to balance structured and unstructured thinking", *Business Strategy Review*, Vol. 14, Issue 2 (Summer 2003), pp. 17-25.

31  Katherine Beatty and Laura Quinn, "Strategic Command: Taking the Long View for Organizational Success", *Leadership In Action*, Vol. 22, No. 2 (May/June 2002), p. 6.

32  Montuori, p. 13.

33  Ronnie Lessem, "Foreword: Requisite Leadership – Managing Complexity", in Elliot Jaques and Stephen D. Clement, *Executive Leadership: A practical guide to managing complexity*, (Falls Church: Cason Hall & Co., 1991), p. xiii.

34  Albert A. Cannella, Jr., "Upper echelons: Donald Hambrick on executives and strategy", Interview, *Academy of Management Executive*, Vol. 15, No. 3 (2001), p. 37.

35  In military circles, the use of detailed plans as a control mechanism is described by the German concept of *befehlstaktik* (detailed control). See: Robert Leonhard, *The Art of Maneuver*, (Novato: Presidio Books, 1991), pp. 52-53.

36  Donald C. Hambrick and Phyllis A. Mason, "Upper Echelons: The Organization as a Reflection of Its Top Managers", *Academy of Management Review*, Vol. 9, No. 2 (1984), pp. 193-206. See also: Kimberly Boal and Robert Hooijberg, "Strategic Leadership Research: Moving On", *Leadership Quarterly*, Vol. 11, No. 4 (2001), pp. 519 and 523, Mason A. Carpenter, Marta A. Geletkanycz and Wm. Gerard Sanders, "Upper Echelons Research Revisited: Antecedents, Elements, and Consequences of Top Management Team Composition", *Journal of Management*, Vol. 30, No. 6 (June 2004), pp. 749-778, Catherine M. Daily, Patricia P. McDougall, Jeffrey G. Covin, and Dan R. Dalton, "Governance and Strategic Leadership in Entrepreneurial Films", *Journal of Management*, Vol. 28, No. 3 (2002), pp. 387-412, and Dusya Vera and Mary Crossan, "Strategic Leadership and Organizational Learning", *Academy of Management Review*, Vol. 29, No. 2 (2004), p. 223.

37  Christopher A. Bartlett and Sumantra Ghoshal, "Changing the Role of Top Management Beyond Systems to People", *Harvard Business Review*, Vol. 73, No. 3 (May-June 1995), pp. 132-142, and Christopher A. Bartlett and Sumantra Ghoshal, "Changing the Role of Top Management: Beyond Strategy to Purpose", *Harvard Business Review*, Vol. 72, No. 6 (November-December 1994), pp. 79-88.

# ENDNOTES

38  Noel M. Tichy and Mary Anne Devanna, *The Transformational Leader*, (Toronto: John Wiley & Sons, 1986), pp. vii, 38, and 222. See also: Robert Heller, *The Super Chiefs*, (Toronto: Penguin Books, 1992), pp. 1 and 12, Rost, p. 8 and Abraham Zaleznik, "The Leadership Gap", *Academy of Management Executive*, Vol. 4, No. 1 (1990), p. 10.

39  Terry Thomas, John R. Schemerhorn Jr., and John W. Dienhart, "Strategic leadership of ethical behavior in business", *Academy of Management Executive*, Vol. 18, No. 2 (2004), p. 56. The Enron scandal refers to the Houston-based energy company that had engaged in fraudulent accounting practices in 2001. WorldCom had engaged in similar activities in 2003 and Andersen had been the auditor for Enron in 2001.

40  Christopher A. Bartlett and Sumantra Ghoshal, "Building Competitive Advantage through People", *MIT Sloan Management Review*, Vol. 43, Issue 2 (Winter 2002), p. 34.

41  Russell Ackoff, "Transformational Leadership", *Strategy & Leadership*, Vol. 27, No. 1 (January-February 1999), pp. 20-25, and Boal and Hooijberg, pp. 526-527, Boal and Hooijberg, p. 525, and House and Aditya, pp. 439-440.

42  Christopher A. Bartlett and Sumantra Ghoshal, "Changing the Role of Top Management: Beyond Structure to Processes", *Harvard Business Review*, Vol. 73, No. 1 (January-February 1995), pp. 86-96.

43  There was a curious correlation between the rise of the business community's empowerment literature in the early to mid-1990s and the rise of manoeuvre warfare advocacy in most western militaries around the same time frame. Empowerment and the German concept of *Auftragstaktik* ('directive control') appear to be based on the same idea of telling subordinates what is desired with the relevant limitations as opposed to how it is to be achieved in detail. The subordinates are then left to achieve the superior's wishes with minimal interference. See: Robert Leonhard, *The Art of Maneuver*, (Novato: Presidio Books, 1991), pp. 50 and 116. Leonhard argued that *Auftragstaktik* represented the exchange of unity of effort and the ability to seize opportunities. See also: William S. Lind, "The Theory and Practice of Maneuver Warfare", in Richard D. Hooker, Ed., *The Maneuver Warfare Anthology*, p. 11 and Franz Uhle-Wettler, "*Auftragstaktik*: Mission Orders and the German Experience", in Hooker, Ed., pp. 236-247.

44  Tichy & Devanna, pp. 5-6, 51, and 216.

45  Finkelstein and Hambrick, pp. 10-11 and Rickards and Clark, pp. 151-152.

46  Frank R. Hunsicker, "Organization Theory for Leaders", in AU-24, Concepts for Air Force Leadership, 4th Edition, (Maxwell AFB: Air University, 2001), pp. 155-157. Astute readers will note that this source is part of the five American Armed Services as opposed to the business community, but the author of the cited article was the chairman of the department of management and marketing at West Georgia College.

47  Hunsicker, pp. 153-154.

48  Mintzberg et al.

49  Robert H. Lowson, "An Introduction to Strategic Management", Strategic Operations Management (2002) p. 42.

50  Henry Mintzberg, "Patterns in Strategy Formation", *Management Science*, Vol. 24, No. 9 (May 1978), p. 941.

51  Henry Mintzberg, "Crafting Strategy", *McKinsey Quarterly*, Issue 3, (Summer 1998), pp. 73-74, 78, 81. See also: Henry Mintzberg and James Waters, "Of Strategies, Deliberate and Emergent",

# ENDNOTES

*Strategic Management Journal*, Vol. 6 (1985), pp. 257-273, Henry Mintzberg, "Strategy Formulation as a Historical Process", *International Studies of Management & Organization*, Vol. 7 Issue 2 (Summer 1977), pp. 28-40, and Burgelman, pp. 64-65.

52  See: Kathleen M. Eisenhardt and L.J. Bourgeois III, "Politics of Strategic Decision-Making in High-Velocity Environments: Towards a Midrange Theory", *Academy of Management Journal*, Vol. 31, No. 4 (1988), pp. 737-770.

53  Detelin S. Elenkov, William Judge and Peter Wright, "Strategic Leadership and Executive Innovation Influence: An International Multi-Cluster Comparative Study", *Strategic Management Journal*, Vol. 26 (2005), p. 667.

54  Elenkov, Judge and Wright, pp. 667-668. Ben L. Kedia, Richard Nordtvedt and Liliana M. Perez, "International Business Strategies, Decision-Making Theories, and Leadership Styles: An Integrated Framework", *CR*, Vol. 12, No. 1 (2002), pp. 42-43.

55  Lieutenant-Colonel Peter Bradley, "Distinguishing the Concepts of Command, Leadership and Management", in Horn and Harris, Eds., p. 115.

56  Burt Nanus, "Leading the Vision Team", *The Futurist*, Vol. 30, No. 3 (May-June 1996), p. 21.

57  W. Glenn Rowe, "Creating wealth in organizations: the role of strategic leadership", *Academy of Management Executive*, Vol. 15, No. 1 (2001), p. 83.

58  Herb Rubinstein, "The Evolution of Leadership in the Workplace", *VISION-The Journal of Business Perspective*, Vol. 9, No. 2 (April-June 2005), p. 48. Italics appear in the original.

59  Bradley, in Horn & Harris, Eds., pp. 109 and 112 and Rost, pp. 79 and 81-82.

60  Stephen R. Covey, *Principle-Centered Leadership*, (Toronto: Summit Books, 1991), pp. 246 and 249, 255. See also: Rost, pp. 106, 150 and 157 and Abraham Zaleznik, "Managers and Leaders: Are They Different?", *Harvard Business Review*, Vol. 55, Issue 3 (May-June 1977), pp. 74-81.

61  Rost, pp. 140-148.

62  Rost refers to this as the 'industrial paradigm of leadership'. See: Rost, pp. 93-94, 134 and 180.

63  For example, see: James R. Meindl, Sanford B. Ehrlich, and Janet M. Dukerich, "The Romance of Leadership, *Administrative Science Quarterly*, Vol. 30, Issue 1 (March 1985), pp. 78-102.

64  See: Leavy, p. 435, Michael Lubatkin and Michael Pitts, "PIMS: Fact or Folklore?", *Journal of Business Strategy*, Vol. 3 Issue 3 (Summer 1985), pp. 38-43, and Sidney Schoeffler, Robert B. Buzzell and Donald F. Heany, "Impact of strategic planning on profit performance", *Harvard Business Review*, Vol. 52, No. 2 (March-April 1974), pp. 137-145.

65  Sidney Finkelstein and Donald Hambrick, *Strategic Leadership: Top Executives and their effects on organizations*, (New York: West Publishing Company, 1996), p. 6.

66  Meryl Davids, "Where Style Meets Substance", *Journal of Business Strategy*, Vol. 16 No. 1 (January-February 1995), pp. 50-51, and Rubinstein, pp. 42-43.

67  See: Ed Kur, "Developing Leadership in Organizations: A Continuum of Choices", *Journal of Management Inquiry*, Vol. 4, No. 2 (June 1995), pp. 198-206.

68  Boal and Hooijberg, pp. 515-517.

69  House and Aditya, pp. 444-445.

70  Albert A. Cannella Jr. and Martin J. Monroe, "Contrasting Perspectives on Strategic Leaders: Toward a More Realistic View of Top Managers", *Journal of Management*, Vol. 23, No. 3 (1997), pp. 213-214.

# ENDNOTES

71. Cannella, "Upper echelons", p. 40.
72. M. Kets de Vries, "Leaders who make a difference", *European Management Journal*, Vol. 14, No. 5 (October 1996), cited in John L. Thompson, "Competence and strategic paradox", *Management Decision*, Vol. 36, No. 4 (1998), p. 274.
73. Harper and Sullivan, p. 44.
74. Rowe, pp. 81-83.
75. Abdalla F. Hagen, Morsheda T. Hassan and Sammy G. Amin, "Critical Strategic Leadership Components: An Empirical Investigation", *SAM Advanced Management Journal*, Vol. 63, No. 3 (Summer 1998), p. 39.
76. Beatty and Quinn, p. 4.
77. "Strategic Leadership in Business", at: http://www.cmoe.com/strategic-leadership.htm. (Sourced on 5 July 2006).
78. R. Duane Ireland and Michael A. Hitt, "Achieving and maintaining the strategic competitiveness in the 21$^{st}$ century: The role of strategic leadership", *Academy of Management Executive*, Vol. 19, No. 4 (2005), p. 63.
79. "Strategic Leadership in Business", at: http://www.cmoe.com/strategic-leadership.htm. (Sourced on 5 July 2006).
80. Clayton M. Christensen, "Making Strategy: Learning By Doing", *Harvard Business Review*, Vol. 75, No. 6 (November-December 1997), pp. 141-156.
81. Bill Richardson, "Comprehensive Approach to Strategic Management: Leading across the Strategic Management Domain", *Management Decision*, Vol. 32, No. 8 (1994), p. 30.
82. Boal and Hooijberg, p. 527.
83. Covey, p. 295. These definitions are close to those offered herein for direct and indirect leadership.
84. Amos Tversky and Daniel Kahneman, "Rational Choice and the Framing of Decisions", *Journal of Business*, Vol. 59, No. 4 (October 1986), p. S257 and Paul C. Nutt, "Framing Strategic Decisions", *Organization Science*, Vol. 9, No. 2 (March-April 1998), p. 195.
85. Richard L. Hughes, "Strategic Leadership", *Leadership in Action*, Vol. 18, No. 4 (1998), pp. 1-6.
86. Rubinstein p. 42.
87. Bruce Pasternak and Albert J. Viscio, "The Centerless Corporation", in Vadim Kotelnikov, "Strategic Leadership", at http://www.1000ventures.com/business_guide/crosscuttings/leadership_strategic.html (Sourced on 5 July 2006).
88. List adapted from John Adair, "Effective Strategic Management", 2002 in Vadim Kotelnikov, "Strategic Leadership", at http://www.1000ventures.com/business_guide/crosscuttings/leadership_strategic.html (Sourced on 5 July 2006).
89. Ireland & Hitt, p. 72.
90. Michael Porter, "What Is Strategy?", *Harvard Business Review*, Vol. 74, No. 6 (November-December 1996), p.77
91. Richard Hughes and Katherine Beatty, "Five Steps to Leading Strategically", *Training & Development*, Vol. 59, No. 12 (December 2005), pp. 46-48.
92. Paul C. Nutt, "Transforming Public Organizations with Strategic Leadership", in A. Halachmi and G. Bouckaert, Eds., *Public Productivity Through Quality and Strategic Management*, (Amsterdam: IOS Press, 1995), p. 80.

# ENDNOTES

93  For example, see: Beatty and Hughes, p. 17.
94  Ireland & Hitt, p. 67.
95  Katherine Beatty and Richard Hughes, "Reformulating Strategic Leadership", *European Business Forum*, Issue 21 (Spring 2005), p. 14.
96  Elenkov, Judge and Wright, p. 669.
97  Richardson, pp. 27-41.
98  Tudor Rickards and Murray Clark, *Dilemmas of Leadership*, (London: Routledge, 2006), p. 147.
99  Barbara J. Davies and Brent Davies, "Strategic Leadership", *School Leadership & Management*, Vol. 24, No. 1 (February 2004), p. 33.
100  Ireland and Hitt, pp. 65-67.
101  Wheatley, pp. 47-48.
102  Minztberg, "Crafting Strategy", p. 76.
103  Cannella & Monroe, pp. 219-224.
104  Cannella & Monroe, pp. 228-230 and Davids, pp. 49-59.
105  For example, see: Lessem, p. xiii-xiv.
106  Beatty and Hughes, p. 16.
107  See: Mason A. Carpenter, Marta A. Geletkanycz and Wm. Gerard Sanders, "Upper Echelons Research Revisited: Antecedents, Elements, and Consequences of Top Management Team Composition", *Journal of Management*, Vol. 30, No. 6 (June 2004), pp. 753-759.
108  Jaques and Clement, p. 288.
109  Michael A. Hitt, Barbara W. Keats, Herbert F. Harback and Robert D. Nixon, "Rightsizing: Building and Maintaining Strategic Leadership and Long-Term Competitiveness", *Organizational Dynamics*, Vol. 23, No. 2 (Autumn 1994), p. 30.
110  Albert A. Vicere, "The Strategic Leadership Imperative for Executive Development", *Human Resource Planning*, Vol. 15, Issue 1 (1992), pp. 27-29.
111  Cited in Sylvain Patenaude, "Competencies for Strategic Planning for the Canadian Forces Health Services", M.A. in Leadership & Training thesis, (Victoria: Royal Roads University, 2005), p. 31.

## Chapter 5   *Scylla*: The Five American Armed Services

1  In the myth, Scylla had six heads as opposed to five.
2  Joint Warfare of the Armed Forces of the United States, Joint Publication 1 (Washington, DC: Joint Chiefs of Staff, 2000), pp. I-1 to I-9 and IV-3 to IV-6, and AFDD 1, pp. 12-13.
3  See p. 18 above.
4  Echevarria, p. 7.
5  Colonel Stephen A. Shambach, US Army, Ed., *Strategic Leadership Primer*, (Carlisle: U.S. Army War College, 2004), p. 6.
6  Betts, p. 37.
7  FM 3.0 *Operations*, (Washington, DC: Secretary of the Army, 2001), Paragraph 2-4.
8  See FM 3.0, Paragraphs 2-5 to 2-14 for the U.S. Army's concepts of the operational and tactical levels of war. The bold typeface appears in the original.

# ENDNOTES

9. It should be noted that each of the five Armed Services are mandated to organize, equip and train their forces. This is not exclusive to the U.S. Army.
10. *Air Force Basic Doctrine*, US Air Force Doctrine Document 1 (AFDD 1), (Washington, DC: Secretary of the Air Force, 2003), p. 11.
11. Figure 1.1, *Organization and Employment of Aerospace Power*, US Air Force Doctrine Document 2 (AFDD 2), (Washington, DC: Secretary of the Air Force, 1998), p. 11.
12. Chapter 2, *Naval Warfare*, Naval Doctrine Publication 1 (NDP 1), (Washington, DC: Secretary of the Navy, 1994).
13. Chapter 2, NDP 1.
14. *Strategy*, MCDP 1-1, (Washington, DC: Secretary of the Navy, 1997), pp. 37-60.
15. MCDP 1-1, p. 53.
16. MCDP 1-1, p. 55.
17. MCDP 1-1, p. 57. See p. 14 above.
18. *U.S. Coast Guard Regulations*, COMDTINST M5000.3B, (Washington, DC: U.S. Coast Guard, 1992), p. 13-1.
19. Lord, pp. 139-145.
20. Betts, p. 40.
21. George McAleer, "Leaders in Transition: Advice from Colin Powell and Other Strategic Thinkers", *Military Psychology*, Vol. 15, No. 4 (2003), pp. pp. 313-314, and Shambach, Ed., p. 3.
22. Department of Strategic Decision Making and Executive Information Systems, Industrial College of the Armed Forces, National Defense University, "Strategic Leader Performance Requirements", *Strategic Leadership and Decision-Making*, http://www.ndu.edu/inss/books/books%20-%201999/Strategic%20Leadership%20and%20Decision-making%20-%20Feb%20 99/pt1ch6.html, p. 7, sourced on 10 November 2006.
23. Strategic Leadership and Decision Making, Part One, p. 2.
24. Colonel W. Michael Guillot, USAF, "Strategic Leadership: Defining the Challenge", *Air & Space Power Journal*, Winter 2003, pp. 70-72 and Shambach, Ed., pp. 12-13.
25. Major Christopher D. Kolenda, US Army, "Transforming How We Fight: A Conceptual Approach", *Naval War College Review*, Vol. LVI, No. 2 (Spring 2003), pp. 100-122. See also: Alan Beyerchen, "Clausewitz, Nonlinearity, and the Unpredictability of War", *International Security*, Vol. 17, No. 3 (Winter 1992/1993), pp. 59-90, Major Glenn E. James, US Air Force, *Chaos Theory: The Essentials for Military Applications*, Newport Paper 10, (Newport: Naval War College, 1996) and James Moffat, *Complexity Theory and Network Centric Warfare*, (Washington, DC: Department of Defense Command and Control Research Program, 2003). Linda P. Beckerman, *The Non-Linear Dynamics of War*, http://www.belisarius.com/modern_business_strategy/beckerman/non_linear.htm, provides a bridge between the five American Armed Services and the business community in this regard.
26. Colonel Michael Flowers, U.S. Army, "Improving Strategic Leadership", *Military Review*, Vol. 84, No. 2 (March-April 2004), pp. 40-41.
27. Colonel Christopher R. Paparone, U.S. Army, "Deconstructing Army Leadership", *Military Review*, Vol. 84 No. 1 (January-February 2004), p. 5.
28. Leonard Wong, *Stifled Innovation? Developing Tomorrow's Leaders Today*, (Carlisle: SSI, 2002), p. 27.

# ENDNOTES

29  Shambach, Ed., p. 14.
30  McAleer, pp. 319-320, and Shambach, Ed., p. 6.
31  This poses a significant challenge for strategic leaders.
32  The Department of Defense dictionary defines the strategic level of war as: "the level of war at which a nation, often as a member of a group of nations, determines national or multinational (alliance or coalition) strategic security objectives and guidance, and develops and uses national resources to achieve these objectives. Activities at this level establish national and multinational military objectives; sequence initiatives; define limits and assess risks for the use of military and other instruments of national power; develop global plans or theater war plans to achieve those objectives; and provide military forces and other capabilities in accordance with strategic plans." See: http://www.dtic.mil/doctrine/jel/doddict/, sourced on 10 February 2007.
33  http://www.dtic.mil/doctrine/jel/doddict/data/c/01089.html (sourced on 6 November 2006).
34  *Army Leadership: Be, Know, Do*, FM 22-100, (Washington, DC: Department of the Army, 1999), p. 1-4.
35  Paparone, "Deconstructing", p. 4.
36  These definitions come from the Department of Defense Joint Dictionary, http://www.dtic.mil/doctrine/jel/doddict/, sourced on 10 February 2007. Note that there are no official Department of Defense terms for leadership and management, but most of the armed services appear to accept a definition where leadership includes both influence and direction. See also: Roy, p. 23 for a similar depiction.
37  George R. Mastroianni, "Occupations, Cultures and Leadership in the Army and Air Force", *Parameters*, Vol. 35, No. 4 (Winter 2005-2006), pp. 76-90.
38  *Leadership and Force Development*, US Air Force Doctrine Document 1-1 (AFDD 1-1), (Washington, DC: Secretary of the Air Force, 2006), p. 1.
39  AFDD 1-1, p. 9. These are roughly aligned with the Army's concepts of 'Direct', 'General' and 'Strategic' leadership.
40  AFDD 1-1, p. 10.
41  Sourced from http://www.navyreading.navy.mil/, 9 November 2006.
42  Shannon A. Brown, "The Sources of Leadership Doctrine in the Air Force", *Air and Space Power Journal* (Winter 2002), pp. 1-11. This is a potential avenue for research for the other American armed services and the CF.
43  Commander Richard N. Rosene, U.S.N.R., "Naval Leadership Assessment and Development", USAWC Strategy Research Project, (Carlisle Barracks: U.S. Army War College, 2005), p. 10.
44  *Leadership Development Framework*, Commandant Instruction M531.3, COMDTINST M5351.3, 9 May 2006, p. 1-1.
45  COMDTINST M5351.3, p. 2-1.
46  See: COMDTINST M5351.3, pp. 3-1 to 3-28.
47  For a discussion of the Afghan Model, see: Stephen Biddle, "Afghanistan and the Future of Warfare", *Foreign Affairs*, Vol. 82, No. 2 (March-April 2003), pp. 31-46 and Echevarria, p. 8.
48  FM 22-100, pp. 7-1 to 7-28.
49  Guillot, p. 67.
50  Guillot, p. 74 and Shambach, Ed., p. 20.

# ENDNOTES

51 Major General Richard A. Chilcoat, *Strategic Art: The New Discipline for 21st Century Leaders*, (Carlisle: U.S. Army War College, 1995), p. 3.

52 Chilcoat, p. 1.

53 Chilcoat, pp. 7-8 and 10. Thucydides was the Ancient Greek historian who wrote *The History of the Peloponnesian War*. Sun Tzu was an Ancient Chinese general who wrote *The Art of War*. General George S. Patton commanded the Seventh US Army in Italy and the Third US Army in Northwest Europe during the Second World War. Field Marshal Erwin Rommel was a German Divisional Commander in France, a Corps Commander in North Africa and an Army Group commander in Italy and North West Europe during the Second World War. General Matthew Ridgeway commanded US XVIII Airborne Corps at Arnhem and the 'Battle of the Bulge', all United Nations ground forces during the Korean War and was Supreme Allied Commander Europe after General Eisenhower. The latter was Supreme Allied Commander Europe during the Second World War and later President of the United States. General George Marshall was the Chief of Staff of the US Army during the Second World War, and later Secretary of State and Secretary of Defense. During his time as the Secretary of State, he was architect of the plan for European reconstruction. Sir Winston Churchill served as the Prime Minister of the United Kingdom during the Second World War.

54 T. Owen Jacobs and Philip Lewis, "Leadership Requirements in Stratified Systems", James G. Hunt and Robert L. Phillips, Eds., *Strategic Leadership: A Multiorganizational-Level Perspective*, (Westport: Quorum Books, 1992), p. 24. 'Direct' leadership, in this case is associated with small groups (from two to a thousand) and 'general' leadership is associated with larger groups.

55 Guillot, p. 72.

56 Major Lista M. Benson, USAF, *Leadership Behaviours at Air War College*, Research Report AU/ACSC/024/1998-04, (Maxwell AFB: Air University, 1998) pp. 13-14.

57 T. Owen Jacobs and Philip Lewis, "Leadership Requirements in Stratified Systems", in Hunt and Phillips, Eds., p. 17.

58 See: James G. Hunt and Robert L. Phillips, "Strategic Leadership: An Introduction", in Hunt and Phillips, Eds., p. 5, and Paparone, p. 3.

59 SLDM, Part One, p. 5.

60 Kimberly B. Boal and Carlton J. Whitehead, "A Critique and Extension of the Stratified Systems Theory Perspective", in Hunt and Phillips, Ed., pp. 240-241. SST even appears in CF doctrine. See: *Leadership In The Canadian Forces: Conceptual Foundations*, (Kingston: CFLI, 2004), Glossary, http://cda-acd.mil.ca/CFLI/engraph/leadership/conceptual/glossary_e.asp, sourced on 17 February 2007.

61 For a discussion of Strategy-Structure-Performance, see p. 35 above.

62 General Charles C. Krulak, USMC, "Strategic Corporal", *Leatherneck*, Vol. 28, Issue 1 (January 1999), pp. 14-18, and General Charles C. Krulak, USMC, "Strategic Corporal: Leadership in the three block war", *Marine Corps Gazette*, Vol. 83, No. 1 (January 1999), pp. 18-23. The original speech was: General Charles C. Krulak, USMC, "The Three Block War: Fighting in Urban Areas", *Vital Speeches of the Day*, Vol. 64, Issue 5 (December 1997), pp. 139-142. The Canadian Army adopted the 'three block war' as a concept to inform its doctrine as the concept offered a simple heuristic device to explain the complexity of contemporary war caused by a significant level of media scrutiny on the battlefield, the challenges posed by the urban environment and

political constraints on the conduct of warfare. The danger with the term is that some may take the '3 blocks' literally as opposed to having to be capable of shifting between three types of operations within a proximate yet flexible time and space.

63. Guillot, pp. 70-71.
64. Simon King, "Strategic Corporal or Tactical Colonel? Anchoring the Right Variable", *Defense & Security Analysis*, Vol. 19, No. 2 (March 2003), pp. 189-190.
65. Thomas J. Williams, "Strategic Leader Readiness and Competencies for Asymmetric Warfare", *Parameters*, Vol. 33 No. 2 (Summer 2003), pp. 23-30.
66. Shambach, Ed., pp. 37-43.
67. McAleer, pp. 312 and 314-316.
68. Walter F. Ulmer, Jr., "Military Leadership into the 21st Century : Another 'Bridge Too Far?'", *Parameters*, Vol. XXVIII No. 1 (Spring 1998), p. 7.
69. Jon P. Briscoe and Douglas T. Hall, "Grooming and Picking Leaders Using Competency Frameworks: Do They Work? An Alternative Approach and New Guidelines for Practice," *Organizational Dynamics,* Vol. 28, No. 2 (Autumn 1999), pp. 48-49, cited in Leonard Wong, Stephen Gerras, William Kidd, Robert Pricone, and Richard Swengros, *Strategic Leadership Competencies*, (Carlisle: US AWC, 2003), p. 5.
70. Wong, Gerras, Kidd, Pricone, and Swengros.
71. General Robert H. 'Doc' Foglesong, USAF, "Leadership from Flight Level 390", *Air and Space Journal*, Vol. XIX, No. 1 (Fall 2004), pp. 8-9.
72. Colonel Fernando Giancotti, "Strategic Leadership and the Narrow Mind: What We Don't Do Well and Why", AU-24, pp. 187-190.
73. AFDD 1-1, p. 10.
74. A copy of this can be found at: http://www.au.af.mil/au/awc/awcgate/navy/navy-ldr-comp.htm (sourced on 13 November 2006).
75. A copy of this can be found at: http://www.au.af.mil/au/awc/awcgate/usmc/leadership.htm (sourced on 13 November 2006).
76. COMDTINST M531.1, p. 2-2.
77. SLDM, "Strategic Leader Performance Requirements", pp. 7-8.
78. Hunsicker, AU-24, p. 158. See also: William E. Turcotte, "Executive Strategy Issues for Very Large Organizations", AU-24, pp. 159-160.
79. Shambach, Ed., p. 13.
80. Shambach, pp. 2-3.
81. McAleer, pp. 318-319.
82. Frank Pagano, "Strategic Leadership", Letter to the Editor, *Military Review*, Vol. 34 No. 6 (November-December 2004), p. 81.
83. Flowers, p. 41 and Ulmer, pp. 9-11.
84. *OFFICER PROFESSIONAL MILITARY EDUCATION POLICY (OPMEP)*, Chairman of the Joint Chiefs of Staff Instruction 1800.01C, (CJCSI 1800.01C), 22 December 2005.
85. See: RADM J.L. Shuford, USN, "President's Forum", *Naval War College* Review, Vol. 59, No. 2 (Spring 2006), pp. 1-2 and College of Naval Warfare, Syllabus for National Security Decision Making, (Newport: Naval War College, 2005), pp. B-5, B-15 to B-16 and P-25.

# ENDNOTES

86    Colonel Mark A. McGuire, U.S. Army, "Senior Officers and Strategic Leader Development", *JFQ: Joint Force Quarterly*, Issue 29 (Autumn/Winter 2001-02), pp. 91-92.

## Chapter 6    Comparing *Scylla* and *Charybdis*

1    See p. 14 for the OCD definitions.
2    Military readers will be reminded of the aphorism "Time spent in reconnaissance is seldom wasted, but never regained".

## Chapter 7    *Odysseus*: The Canadian Forces

1    Homer, p. 163.
2    Howard Coombs with Richard Goette, "Supporting the Pax Americana: Canada's Military and the Cold War", in Colonel Bernd Horn, Ed., *The Canadian Way of War*, (Toronto: Dundurn Press, 2006), p. 268.
3    Government of Canada, *A Role of Pride and Influence in the World*, International Policy Statement, (Ottawa: Government of Canada, 2005).
4    This framework comes from Donald Nuechterlein, *America Overcommitted: United States National Interests in the 1980s*, (Lexington: University of Kentucky Press, 1985), pp. 6-30.
5    Andrew Cooper, Richard Higgott and Kim Richard Nossal, *Relocating Middle Powers: Australia and Canada in a Changing World Order*, (Vancouver: UBC Press, 1993), pp. 19 and 116.
6    Andrew Cooper, "Between Fragmentation and Integration: The Evolving Security Discourse in Australia and Canada", *Australian Journal of International Affairs*, Vol. 49, No. 1 (May 1995), p. 53.
7    Ross Graham, "Civil Control of the Canadian Forces: National Direction and National Command", *Canadian Military Journal*, Vol. 3, No. 1 (Spring 2002), p. 25.
8    Major Jeff Tasseron, Canadian Forces, "Facts and Invariants: The Changing Context of Canadian Defence Policy", *Canadian Military Journal*, Vol. 4, No. 2 (Summer 2003), p. 20.
9    Cooper, Higgott and Nossal, pp. 4 and 36.
10    Tasseron, p. 23.
11    Conrad Winn, president of COMPAS, cited in Fen Osler Hampson and Dean Oliver, "Pulpit Diplomacy", *International Journal*, Vol. LIII, No. 3 (Summer 1998), p. 379.
12    Graham, p. 25. See also: Coombs with Goette, in Horn, Ed., pp. 266 and 272 and Scot Robertson, "Years of Innocence and Drift: The Canadian Way of War in the Post-Cold War Era", in Horn, Ed., pp. 360-361.
13    David Haglund, "Here Comes M. Jourdain: A Canadian Grand Strategy Out of Molière", *Canadian Defence Quarterly*, Vol. 27, No. 3 (Spring 1997), pp. 20-21.
14    Hampson and Oliver, p. 403.
15    Lieutenant Colonel John Blaxland, Australian Army, "The Armies of Canada and Australia: Closer Collaboration?", *Canadian Military Journal*, Vol. 3, No. 3 (Autumn 2002), pp. 50-51.
16    See David Eaves, "Central Myth of Canadian Diplomacy", *Toronto Star*, 6 November 2006. Eaves argued that Lester B. Pearson's desire to solve the Suez crisis was motivated not by

# ENDNOTES

   altruism but self-interest as a Middle East war that burgeoned into a superpower exchange of nuclear weapons would inevitably affect Canada.
17 See: Sean M. Maloney, *Canada and UN Peacekeeping: Cold War by Other Means, 1945-1970*, (St. Catharines: Vanwell Publishing Limited, 2002). See also: Coombs with Goette, in Horn, Ed., p. 287.
18 Donna Winslow, "Canadian Society and Its Army", *Canadian Military Journal*, Vol. 4, No. 4 (Winter 2003-2004), p. 12.
19 Coombs with Goette, in Horn, Ed., p. 289.
20 Bentley, pp. 30 and 96.
21 Colonel J.H. Vance, "Tactics without Strategy or Why The Canadian Forces Do Not Campaign", in English et al., pp. 272-273, and 280-281. See also: Joel Sokolsky, "The Politics of Defence Decisions at Century's End", in Horn and Harris, Eds., p. 348. For an example of this stratagem from the Korean War, see: Coombs with Goette, in Horn, Ed., p. 274
22 *Repertory of Army Terminology,* A-GL-397-000/JX-001, (Kingston: Directorate of Army Doctrine, 2006).
23 Paragraph 4, Part I, National Defence Act N-5 (sourced from http://lois.justice.gc.ca/en/N-5/269056.html#rid-269061 on 17 November 2006 on 17 November 2006) and Paragraph 18, Part II, National Defence Act N-5 (sourced from http://lois.justice.gc.ca/en/N-5/269085.html on 17 November 2006).
24 Agency theory represents a form of transactional leadership. For a discussion of agency theory in the US military context, see: Peter D. Feaver, *Armed Servants: Agency, Oversight And Civil-Military Relations*, (Cambridge, MA: Harvard University Press, 2003).
25 *Leadership in the Canadian Forces: Leading the Institution*, (Kingston: Canadian Forces Leadership Institute, 2006), p. 84. (Henceforth *LIM*)
26 *LIM*, p. 66.
27 *LIM*, p. 66 and Thorne, pp. 8-9.
28 *LIM*, p. 48.
29 Major General Douglas Dempster, Canadian Forces, "Generalship and Defence Program Management", in Horn and Harris, Eds., p. 456.
30 *LIM*, pp. 12-13, 19, and 113-114.
31 Jacobs, pp. 51-56, and 131-178.
32 *Leadership In The Canadian Forces: Conceptual Foundations*, (Kingston: CFLI, 2004), Glossary, http://cda-acd.mil.ca/CFLI/engraph/leadership/conceptual/glossary_e.asp, sourced on 9 Feb 07.
33 Definitions for all three terms appear in the Glossary of *Conceptual Foundations*.
34 Bradley, in Horn and Harris, Eds., p. 107 and 109.
35 *Leadership In The Canadian Forces: Leading People* (Kingston: Canadian Forces Leadership Institute, 2006), pp. 33-35.
36 Bradley, in Horn & Harris, Eds., pp. 118-119.
37 For details, see: *Duty with Honour: The Profession of Arms in Canada*, A-PA-005-000/AP-001, (Kingston: Canadian Defence Academy, 2003), pp. 25-34.
38 *Leadership In The Canadian Forces: Leading People*, p. 6. See also: Bentley, p. 121 and *LIM*, p. 72.

# ENDNOTES

39 *Leadership In The Canadian Forces: Conceptual Foundations*, (Kingston: Canadian Forces Leadership Institute, 2004), p. 98.
40 Bentley, p. 122.
41 *Leadership In The Canadian Forces: Doctrine*, pp. 38-39.
42 *Leadership In The Canadian Forces: Conceptual Foundations*, p. 98.
43 *Leadership In The Canadian Forces: Doctrine*, pp. 36-37.
44 *Leadership In The Canadian Forces: Conceptual Foundations*, p. 100.
45 *LIM*, pp. 97-119.
46 For example, see: Colonel Alan J. Howard, Canadian Forces, "Influencing Transformation: Military Leadership at the Strategic Level", NSSC Paper, (Toronto: CFC, 2004), p. 7, and Peskett, in English et al., p. 119.
47 *Leadership In The Canadian Forces: Conceptual Foundations*, p. 118. See also: Vice Admiral Gary Garnett, Canadian Forces, "The Flag and General Officer as a Resource Manager", in Horn and Harris, Eds., p. 467.
48 Robert W. Walker and L. William Bentley, *A Professional Development Framework: Canadian Forces Leadership, Professionalism, Leader Capacities and Leader Development – A Summary*, (Kingston, CFLI, 2006), pp. 2-4.
49 *PD Framework*, p. 6.
50 Bentley, p. 91.
51 Colonel Stuart Beare, Canadian Army, *Operational Leadership Experience In Officer Professional Development: A Pillar In Peril*, (Toronto: CFC, 2000), and Ronald G. Haycock, "The Labours Of Athena And The Muses: Historical And Contemporary Aspects Of Canadian Military Education", *Canadian Military Journal*, Vol. 2 No. 2 (Summer 2001), pp. 5-22.
52 *The Canadian Forces College National Security Studies Course Course 8 Programme Syllabus*, CFC 450, (Toronto: Canadian Forces College, 2005), p. 2-1/1.
53 CFC 450, p. 3-A-1/4.